STATISTICS OF QUALITY CONTROL

SAMPLING INSPECTION AND RELIABILITY

STATISTICS OF QUALITY CONTROL

SAMPLING INSPECTION AND RELIABILITY

Suddhendu Biswas
Retired Professor, Department of Statistics and
Formerly the Head and Dean
Faculty of Mathematical Sciences, University of Delhi

Editorial Assistance

G. L. Sriwastav
Professor in Statistics, Dibrugarh University, Assam

New Central Book Agency (P) Ltd
LONDON
DELHI KOLKATA PUNE HYDERABAD ERNAKULAM

REGD OFFICE
8/1 Chintamoni Das Lane, Kolkata 700 009, India
email: ncbapvtltd@eth.net

OVERSEAS
NCBA (UK) Ltd, 149 Park Avenue North, Northampton, NN3 2HY, UK

EXPORT
NCBA Exports Pvt. Ltd, 212 Shahpur Jat, New Delhi 110 049
email: ncbaexp@ncbapvtltd.com

BRANCHES

208B Peacock Lane, Shahpur Jat
New Delhi 110 049
email: ncbadel@ncbapvtltd.com

Shop Nos. 3 & 4, Vinayak Towers
681/B Budhwar Peth, Appa Balwant Chowk
Pune 411 002
email: ncbapune@ncbapvtltd.com

House No. 3-1-315, 1st Floor
Nimboliadda, Kachiguda, **Hyderabad** 500 027
email: ncbahydb@ncbapvtltd.com

GSS Shopping Complex, 1st Floor
Opposite College Post Office, Convent Road
Ernakulam 682 035
email: ncbaernk@ncbapvtltd.com

STATISTICS OF QUALITY CONTROL

First Published: 1996
Second Revised Central Edition: 2003
Third Edition: 2007
Reprinted: 2011

PUBLISHER AND TYPESETTER
New Central Book Agency (P) Ltd
8/1 Chintamoni Das Lane, Kolkata 700 009

PRINTER
New Central Book Agency (P) Ltd
Web-Offset Division, Dhulagarh, Sankrail, Howrah

COVER DESIGNER
Soumen Paul

COVER PRINTER
Liba Graphics, Kolkata

PROOF CHECKING
Pradip Kumar Chowdhury

PROJECT SUPERVISOR
Abhijit Sen

ISBN: 81-7381-433-3

Price: ₹ 215.00

Dedicated to
my beloved wife
BANI (1936–88)
who has left her pearl
like writing for me to keep her
alive ever in my memory

Preface to the Second Revised Central Edition

The BOOK has considerably been revised and corrected with minor addition and deletion in respect of the original first three chapters in the first edition. A few illustrative examples one on the graphical method of drawing AOQ curve and the other on methods of determining (n, c) of a plan using S-plus programming have been added to facilitate better understanding of the Reader. The inclusion of the variable sampling inspection technique by Indian Standard Institution together with the elaboration of the method of using the tables supplemented by examples is a unique feature of this edition compiled in Appendix A. A completely new chapter under the title of 'Maintenance' has now been added to give a sizeable account of 'Reliability and Maintenance' as covered by syllabus of many of the universities in the course of Statistics, Operation Research and Engineering Reliability.

I remain greatly indebted to my former student Dr G.L. Sriwastav, Professor in Statistics, Dibrugarh University to undertake the great painstaking effort of going through the text very meticulously and suggesting correction of several errors and improvements in the text.

My thanks go also to Mr Amar Gurung (a PhD Research Scholar from Tribhuvan University, Nepal to do his Doctoral programme under my supervision in the University of Delhi), for his constant help.

My thanks to the students of the Department of Applied Operational Research, University of Delhi during the last few academic sessions to suggest many improvements while being readers of this text.

My daughter, Swati, although far away while working as a Teaching Associate in Statistics at the Ohio State University, Columbus, U.S.A. deserves also my appreciation for many of her suggestions. Finally, my thanks to Ms. Rumi Dutta for going through the galley proofs.

July 5, 2002
24 Uttaranchal Apartments
5 I.P. Extension
Delhi 110092 India

Suddhendu Biswas

Preface to the First Edition

The book is based on my lecture notes to Postgraduate students in Statistics in the University of Delhi since about one and a half decade.

The plan is to cover a Mathematically oriented course in Industrial Statistics in a single text with comprehensive style and easy readability for students with elementary background of Algebra, Calculus and Probability. Although, the author lacks background of any involvement with applied project in Industry, he hopes that the text written on a theoretical plane may still help an average practitioner in the field with some background of Mathematics or Statistics.

The book is divided into three broad parts. The first part deals with sampling inspection techniques. A critical appraisal of lot and continuous sampling plans (for both attributes and variables) have been appended with large number of examples and counter-examples; especially, non-efficacies of usual lot sampling plan with Binomial process curve, etc. have been focussed. Curtailed sampling plan (evolved by Columbia University research group) and Semi-curtailed variants of the same is also a new item included in the book. Other critical appraisals not included in a text of this form are inadequacy of Neyman-Pearsonian procedure in the sampling inspection plan and almost universal usage of O.C. function rather than power function in the sampling inspection context. The second part is devoted to control charts and other related techniques used in quality control. A full discussion of Tolerance Range (both Normal as well as based on Ordered Statistics) showing its applicability in several quality control problems and its theoretical superiority over control charts is another unique feature of the book. Besides, the Economic control charts using Duncan's the CUSUM charts for testing the process in control (or not) by using graphical V-mask technique, decision interval techniques and the associated distribution of average run length have all believed to have made, the text exhaustive for the study of quality control. A large number of examples have been sprinkled to stimulate better comprehension into the fundamentals of the subject. Finally, the last chapter has been devoted to the theory of Reliability using the techniques of Stochastic processes. Here also, almost all possible systems have been analysed with a large number of illustrations and solved examples. A part from some complicated Reliability models like mixed systems with limited repairability, special topics like Preventive Maintenance Scheme and coherent systems have also been discussed, of course, within the limitations of the text. A detailed appendix providing the background of the Theory of Laplace transforms (which has been abundantly used throughout the study of Reliability) and Renewal theory (which is prerequisite for the study of Reliability) has been appended at the end of the text for the convenience of the readers.

The manuscript for the book was written in picturesque handwriting by my late wife Bani along with her sister. My wife is no longer here to see that the book has been the light of the day! However, my sister-in-law Kalyani will indeed be happy to see that her efforts have not gone in vain. I deeply acknowledge her write-up inclusive of artistic sketches of several figures in the text.

It is difficult to forget the great stalwarts who profoundly influenced my thoughts in all my academic pursuits. No less than later Professor VS. Huzurbazar (of University of Poona and lately of University of Denver), my preceptor, guide and philosopher. Professor Badrig. M. Kurkjian (lately of the University of Alabama) whose golden human touch and encouragement will ever remain with me; and Mr Francis Sutton (Director, Rockefeller Bellagio Residency Programme, Italy) who almost hypnotised me by his wonderful versatility (even in Mathematical Sciences), eloquency and ever stretching friendly helping hand. His wonderful piano recitals during my lonely days in Bellagio has gone deep in my heart. Last, but not least to Dr (Ms) Donna Marewick of the University of Melbourne, Australia for making me conscious of my ability which I do not deserve.

I remain grateful to all my colleagues in the Department of Statistics in the University of Delhi for their ungrudging support and cooperation in all my endeavours. Professor B.B. Chakraborty of the Department of Mathematics requires special mention. My great thanks to my students Dr (Mrs) Mariamma Thomas, Dr Monthir A. Abid, Professor Abul Basher Mian, Dr Tapan Kumar Pachal, Dr Shishir Kumar Jha, Dr (Mrs) Sunita Chitkara, Dr (Mrs) Rita Jain, Dr Vijay Kumar Sehgal, Dr (Mrs) Gurprit Grover, Dr Saroj Kumar Adhikari, Mrs Jashvinder and Dr Debasree Goswami.

My appreciation to my daughter Swati who helped me in several computations in the text.

August 31, 1996 Suddhendu Biswas

Department of Statistics
University of Delhi, Delhi

Contents

Chapter 1

Sampling Inspection Plan in Quality Control

1.1 Introduction on Product Control

Inspection for acceptance purposes are carried out at many stages in the manufacturing process. This may be inspection of the incoming materials; process inspection at various points in the manufacturing process; and finally inspection of the manufactured products by the producer, before the same is being marketed. We are, however, concerned with the inspection by consumers (i.e., purchasers).

With the completion of production of manufacturing items (while maintaining a certain standard or quality by adoption of suitable quality control techniques), it becomes necessary to make the entire production marketable or acceptable to the consumer. The process, therefore, starts with interaction between producer and consumer. While producer guarantees a certain level of quality to make his product marketable, the consumer, whose interest primarily lies in the quality of his purchase, may not accept in toto the producer's certificate in respect of the standard or quality of products. He likes to examine the same, whereby the producer is obliged to allow him to undergo the testing process. While cent per cent checking is not physically or practically possible, one has to resort to sampling inspection, ultimately. This, of course, brings certain degree of uncertainties in the process of selection. A good quality of produced materials may be rejected on the basis of a few sample items which brings some risk to the producer. On the other hand, bad quality or bad lots of manufactured products are accepted on the basis of a few favourable or good items drawn in the sample. This, again introduces some degree of risk to the consumer. Notwithstanding this problem of sample selection, one cannot afford to undergo complete enumeration or cent per cent testing because of the following reasons:

- Testing each and every item in a lot involves high cost. Therefore, it would be quite uneconomical if one has to pay very high prices for simply finding whether the lot is acceptable or not.
- The time required to test, *say*, 1000 items in a lot, may also be quite long which, neither producer nor consumer can afford to waste simply to come to a decision.
- The testing procedure may often be destructive, (*say*, the testing of manufactured bulbs) which means the cost of testing will be the cost of all the manufactured items which is, of course, very ridiculous.

Therefore, it would only be practicable, if a lot of large number of manufactured items could be accepted (or rejected) on the basis of the testing of a reasonably small number of samples randomly drawn from the population of every such lot of large number of manufactured items. That is, with every lot of, *say,* N number of items, a random sample of $n(n < N)$ is drawn and a decision in respect of the acceptance of the lot of size N is thereby made only on the basis of the inspection of n number of items. If our concern is the attribute (i.e., we call an item to be defective, if it fails to satisfy one or more of the desired criteria), then we accept a lot only when the number of defective items in the sample does not exceed a preassigned number known as *acceptance number* (*say,* c out of n). On the other hand, if the number of defective items, *say,* r, exceeds c (the acceptance number), then we take a decision to reject the lot. Needless to add, that a sampling inspection plan as above one may still involve certain degree of risk or uncertainty in the sale-purchase process and the sale or purchase of bad lot of manufactured items are never totally precluded in the sampling inspection plan. However, by keeping producer's and consumer's risk at more or less the same level, one can show that such sampling inspection plans not only save cost, but also remains equally fair to both consumer and producer.

Finally, while noting the unavoidability of *Producer's Risk* (PR) and *Consumer's Risk* (CR) in any sampling inspection plan, it is necessary to put both of them at a fairly low level, as far as possible at the same level, to maintain the same level of fairness to both consumer and producer. The objective of the producer is to ensure that the sampling plan has a low probability of rejecting good lots. Lots are defined as *high quality* if that contain no more than a specified proportion of the defective known as *Acceptable Quality Level* (AQL). Again the objective of the consumer is to ensure that the sampling plan has a low probability of accepting bad lots. Lots are defined as bad quality if the proportion of defective materials is greater than a specified amount known as *Lot Tolerance Percentage Defective* (LTPD).

1.2 Consumer's Risk and Producer's Risk in Single Sampling Plan

Notations:

N ≡ lot size

n ≡ size of the random sample from each lot

r ≡ no. of the defective in a random sample of size n

c ≡ acceptance number (i.e., if $r > c$, then the lot will be rejected on the basis of sample inspection)

p ≡ true proportion of the defective in the lot. p_p and p_c are the producer's and consumer's value of p.

Then assuming the sampling distribution of the number of defective materials to be binomial with parameters (n, p), we have

$$\text{PR} = 1 - \sum_{r=0}^{c} ({}^nC_r) p_p^r (1 - p_p)^{n-r}, \text{ where } p_p < c/n \tag{1.2.1}$$

$$\text{and CR} = \sum_{r=0}^{c} ({}^nC_r) p_c^r (1 - p_c)^{n-r}, \text{ where } p_c > c/n. \tag{1.2.2}$$

In case, the sampling distribution of the number of defective in Poisson with mean number of defective to be np, then

$$\text{PR} = 1 - \sum_{r=0}^{c} \frac{e^{-np_p}(np_p)^r}{r!}, \text{ where } p_p \leq c/n \tag{1.2.3}$$

$$\text{CR} = \sum_{r=0}^{c} \frac{e^{-np_c}(np_c)^r}{r!}, \text{ where } p_c > c/n. \tag{1.2.4}$$

Again PR and CR can be taken as type I and type II error in sample testing of hypothesis

$$H_0 : p = p_p \leq c/n \text{ against } H_1 : pp_c > c/n, \tag{1.2.5}$$

where p_p and p_c are the values of p given by the producer and consumer respectively. The critical region w is given by

$$w : r > c \tag{1.2.6}$$

and the acceptance region $(W - w)$ is given by

$$W - w : 0 \leq r \leq c. \tag{1.2.7}$$

Now the type I error of $E = P(E \in w \mid H_0)$, where we define $E \equiv (x_1, x_2, \ldots, x_n)$ as a sample point. We define a random variable (r.v.) x_i such that

$$\begin{aligned} x_i &= 1, \text{ when the item is defective} \\ \text{and } x_i &= 0, \text{ otherwise} \end{aligned}$$

then

$$\sum_{i=1}^{n} x_i = r.$$

We define error of type I as

$$E_1 = \alpha = P(E \in w \mid H_0) = P\left(\sum_{i=1}^{n} x_i = r > c \mid H_0\right) = \text{PR} \tag{1.2.8}$$

and the error of type II as

$$E_2 = \beta = P(E \in W - w \mid H_1) = P\left(\sum_{i=1}^{n} x_i = r \leq c \mid H_1\right) = \text{CR}. \tag{1.2.9}$$

On the other hand, if $H_0 = p = p_2 > c/n$ is tested against $p = p_1 \leq c/n$, then E_1 and E_2 are CR and PR respectively.

1.3 A Remark on the Applicability of Neyman-Pearsonian Testing of Hypothesis in Sampling Inspection Plan

The Neyman-Pearsonian (NP) testing of hypothesis against an alternative hypothesis is based on the principle of keeping E_1 fixed and minimising E_2 subject to fixed level of E_1. This implies that we keep PR (or CR) at fixed level depending on $H_0 :\ p = p_1$ (or $p = p_2$) against $H_1 :\ p = p_2$ (or $p = p_1$) and minimising CR (or PR). This, however, may lead to a disparity to the value of CR and PR which may not be workable in a sample inspection plan. On the other hand, if we keep both CR and PR at preassigned low but almost equal level and the sample size, a random variable, then the testing procedure may be more amenable to producer and consumer.

1.4 Operating Characteristic Function

In what follows from the discussion of the earlier section, the probability that a lot of size N with proportion of the defective p will be accepted is

$$P(p) = \sum_{r=0}^{c} ({}^nC_r) p^r (1-p)^{n-r}, \tag{1.4.1}$$

where c is the Acceptance Number (AN), if the sampling distribution of the number of defective is $B(n, p)$. Given c, $P(p)$ in 1.4.1 is called the OC *function of the test,* if we plot $P(p)$ for different values of $p (0 \leq p \leq 1)$, we get what is known as *Operating Characteristic curve* or OC *function.* Thus, it represents the probability that the sample point E falls in the acceptance region $(W - w)$. Again this is

$$P(p) = 1 - P(E \in w) = 1 - \text{power function}, \tag{1.4.2}$$

where $P(E \in w)$ is the power function. For any given sampling distribution plan (n, c), OC curve can be obtained readily. A typical form of OC curve is given in Fig 1.

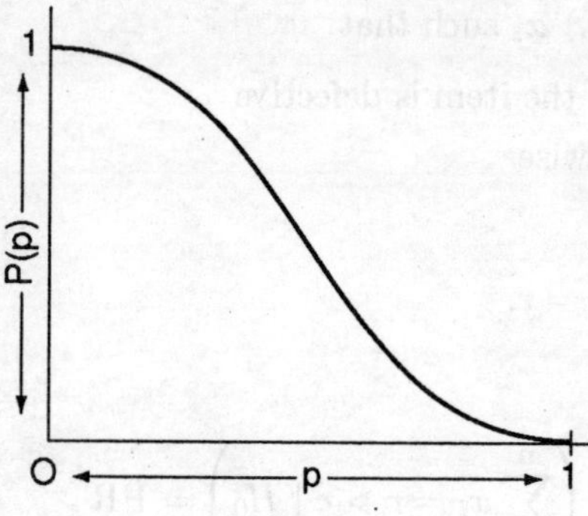

Fig 1

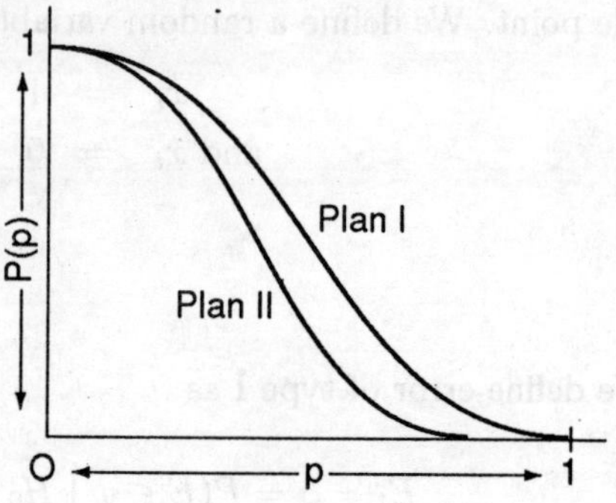

Fig 2

Let us consider two sampling plans, *say,* (n_1, c_1) and (n_2, c_2). How do you find out which plan is better? Given two plans we can have OC curves of the plans. Let us assume that it conforms to the following curves as in Fig 2.

Obviously, the plan II is better, because, for lower values of p (i.e., when quality is maintained) it gives a higher probability of acceptance; whereas, for higher values of p (when quality or standard is low) it gives a much lower probability of acceptance compared to plan I. In other words, plan II corresponds to OC function which is more sensitive to quality which is characterized by its steepness. Thus, steeper is the OC function, better is the plan.

An ideal OC function is given by

$$\begin{aligned} (i)\ P(p) &= 1 \text{ if } p \leq p_0 \\ &= 0 \text{ if } p > p_0 \end{aligned} \tag{1.4.3}$$

$$\begin{aligned} \text{or, } (ii)\ P(p) &= 1 \text{ if } p \leq p_0 \\ &= 0 \text{ if } p > p_1 \end{aligned} \tag{1.4.4}$$

leaving (p_0, p_1) region as indifference region, i.e., we do not mind whether we accept or reject when $p > p_0$ but $p \leq p_1$ given in Fig 3. Unfortunately, neither of the plans can be materialized unless we make 100% inspection.

However, it is possible to approximate the second function as ideally conceived function by slight approximation. That is, we make $P(p)$ to vary near to unity, *say,* $P(p) = 1 - \alpha$ at some value p_0 of p and $P(p)$ is very near to zero; $P(p) = \beta$ at some value p_1 of p, $p_0 < p_1$. Now the specification can be achieved; i.e., we have now four points to draw the OC curve, viz.,

$$P(0) = 1,\ P(p_0) = 1 - \alpha,\ P(p_1) = \beta,\ P(1) = 0$$

from which OC function can be drawn as shown in Fig 3.

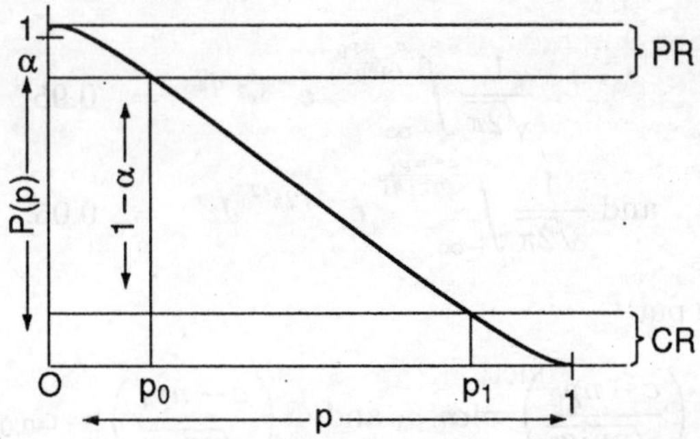

Fig 3

Remarks

As discussed above Neyman-Pearsonian plan is not suitable for testing in sample inspection plan. All what is desirable is equivalent CR and PR; which leads the sample size to be a random variable as in the case of Sequential Probability Ratio Test (SPRT) of Wald. OC function is more frequently used here than power function. In this case, because of basic difference in the sample inspection plan (viz., e_1 and e_2 are fixed and sample size being a random variable), the OC function $\neq$ [1– power function (of fixed sample size)]. However, the OC function in case of SPRT is more readily available, like power function in NP test.

Even there may be several OC functions within the specifications. The choice of the best OC function is guided by the sensitivity (or steepness) of the OC function towards quality.

Example 1 *Suppose* α, β, p_0 *and* p_1 *are given by* $\alpha = \beta = 0.05$; $p_0 = 0.01$ *and* $p_1 = 0.04$. *Then explain how you would get the plan* (n, c). *Suggest a method of getting* (n, c) *under normal approximation to binomial.*

Solution: We have

$$\sum_{r=0}^{c} ({}^nC_r)(0.01)^r(1 - 0.01)^{n-r} = 0.95 \qquad \text{(i)}$$

and

$$\sum_{r=0}^{c} ({}^nC_r)(0.04)^r(1 - 0.04)^{n-r} = 0.05. \qquad \text{(ii)}$$

Under normal approximation (*i*) and (*ii*) correspond to (*iii*) and (*iv*) respectively and are given by

$$\frac{1}{\sqrt{2\pi n p_0 q_0}} \int_{-\infty}^{c} \exp - \frac{(x - np_0)^2}{e^{2(np_0q_0)}}\, dx = 0.95 \qquad \text{(iii)}$$

and

$$\frac{1}{\sqrt{2\pi n p_1 q_1}} \int_{-\infty}^{c} \exp - \frac{(x - np_1)^2}{e^{2(np_1 q_1)}}\, dx = 0.05. \qquad \text{(iv)}$$

Putting

$$\frac{x - np_0}{\sqrt{np_0 q_0}} = z \text{ and } \frac{x - np_1}{\sqrt{np_1 q_1}} = z',$$

we have

$$\frac{1}{\sqrt{2\pi}} \int_{-\infty}^{\frac{c-np_0}{\sqrt{np_0 q_0}}} e^{-\frac{1}{2z^2}\, dz} = 0.95$$

$$\text{and } \frac{1}{\sqrt{2\pi}} \int_{-\infty}^{\frac{c-np_1}{\sqrt{np_1 q_1}}} e^{-1/2z'^2}\, dz' = 0.05.$$

In the next place, we put

$$\Phi\left(\frac{c - np_0}{\sqrt{np_0 q_0}}\right) = \alpha_{0.95} \text{ and } \Phi\left(\frac{c - np_1}{\sqrt{np_1 q_1}}\right) = \alpha_{0.05}. \qquad (1.4.5)$$

Hence we can solve (n, c) from the above, where

$$\frac{1}{\sqrt{2\pi}} \int_{-\infty}^{t} e^{-1/2z^2}\, dz = \Phi(t)$$

and $\alpha_{0.95}$ and $\alpha_{0.05}$ are 95% and 5% critical values of the standard normal deviate, respectively.

Remarks

The method based on approximation to normality will fail if n is reasonably small.

1.5 Corrective Sampling Plan (CSP)

A modification of the acceptance sampling plan is given by corrective sampling plan (CSP). There are two variants of the same, viz., CSP with or without replacement.

The basic idea is that the consumer is obliged to accept the lot even if the lot is otherwise not acceptable by the principle of acceptance sampling plan, described earlier. This is done by making a thorough inspection of the residual $(N - n)$ items and replacing the defective by good items (in case of sampling with replacement) or throwing away bad items without exchange of the same by good items (in case of sampling without replacement). Obviously, the cost of inspection in corrective sampling plan is far more than acceptance sampling plan.

1.6 Average Sample Number (ASN)

OC function gives an idea about the quality of the sampling inspection plan. But the feasibility of a very good plan is sometimes restricted by the high cost of the plan. The cost of a plan depends upon the average sample number (ASN). Thus the cost of plan = ASN × cost per unit of inspection.

In case of an acceptance sampling plan

$$\text{ASN} = I = n, \tag{1.6.1}$$

whereas for a corrective sampling plan

$$\text{ASN} = n + [1 - P(p)](N - n), \tag{1.6.2}$$

where $P(p)$ = Probability of acceptance.

1.7 Average Outgoing Quality (AOQ)

AOQ measures the quality of the lots measured by the proportion of defectives on the average in the lots. Obviously, AOQ = p, in acceptance sampling plan with stable level of quality of production. However, for corrective sampling plan, AOQ with replacement is given by

$$(\text{AOQ})_{\text{rep}} = \frac{Np - Ip}{N} \tag{1.7.1}$$

and without replacement as

$$(\text{AOQ})_{\text{w.rep}} = \frac{Np - Ip}{N - Ip}. \tag{1.7.2}$$

Again

$$(\text{AOQ})_{\text{rep}} = \frac{Np - np - [1 - P(p)](N - n)p}{N},$$

where $I = n + [1 - P(p)](N - n)$, $P(p)$ being the probability of acceptance.

$$\therefore\ (\text{AOQ})_{\text{rep}} = \frac{(N - n)p[1 - [1 - P(p)]}{N} = \frac{(N - n)pP(p)}{N} \tag{1.7.3}$$

$$= \left(1 - \frac{n}{N}\right) pP(p) = pP(p), \text{ when } \frac{n}{N} \to 0. \tag{1.7.4}$$

Similarly, AOQ without replacement is given by,

$$(\text{AOQ})_{\text{w.rep}} = \frac{(N - n)pP(p)}{(N - np) - [1 - P(p)](N - n)p} \tag{1.7.5}$$

In either of the cases, it may be seen that AOQ vanishes at the points $p = 0$ (i.e., when the lot is accepted with probability unity) and $p = 1$ [i.e., when the lot is rejected with probability of acceptance, $P(a) = 0$], subject to the condition that the same lot is accepted with complete replacement (or throwing away) of the defective items by good ones.

►► NOTE

For perfect quality, AOQ = 0 and the lot is not accepted in the latter case if it does not make any change in the quality of the incoming units in the accepted lot otherwise.

1.8 Graphical Method of Drawing AOQ Curve

Using (1.7.4) and the AOQ for a binomial process curve with replacement is

$$\begin{aligned}(\text{AOQ})_{\text{rep}} &= \frac{p(N-n)}{N}\sum_{r=0}^{c}\binom{n}{r}p^r(1-p)^{n-r} \\ &= \frac{p(N-n)}{N}\frac{1}{B(n-c,c+1)}\int_0^{1-p}t^{n-c-1}(1-t)^c\,dt. \qquad (1.8.1)\end{aligned}$$

Given N, n and c, one can get AOQ curve for different values of p.

$$= p\frac{N-n}{N}[\text{pbeta}\,(1-p, n-c, c+1)],$$

where

$$\frac{1}{B(n-c,c+1)}\int_0^{1-p}t^{n-c-1}(1-t)^c\,dt$$

is denoted by

$$\text{pbeta}\,(1-p, n-c, c+1). \qquad (1.8.2)$$

Taking $N = 500$, $n = 10$ and $c = 1$ and denoting

$$(\text{AOQ}) = p(N-n)/N[\text{pbeta}\,(1-p, n-c, c+1)], \quad \text{[from (1.8.1) and (1.8.2)]}$$

we compute AOQ curve for different values of p as follows:

```
Working data will be in C:\Program Files\splus4\users\default\_Data
>pbeta(0.99,9,2) for p=0.01
[1] 0.9957338
>pbeta(0.95,9,2) for p=0.05
[1] 0.9138616
>pbeta(0.90,9,2) for p=0.10
[1] 0.7360989
>pbeta(0.85,9,2) for p=0.15
[1] 0.5442998
>pbeta(0.80,9,2) for p=0.20
[1] 0.3758096
>pbeta(0.75,9,2) for p=0.25
[1] 0.2440252
>pbeta(0.70,9,2) for p=0.30
[1] 0.1493083
>pbeta(0.65,9,2) for p=0.35
[1] 0.08595444
>pbeta(0.60,9,2) for p=0.40
[1] 0.0463574
>pbeta(0.55,9,2) for p=0.45
[1] 0.0232571
>pbeta(0.50,9,2) for p=0.50
[1] 0.01074219
```

```
>pbeta(0.40,9,2) for p=0.60
[1] 0.001677722
>pbeta(0.30,9,2) for p=0.70
[1] 0.0001436859
>pbeta(0.20,9,2) for p=0.80
[1] 4.1984e-006
>pbeta(0..10,9,2) for p=0.90
[1] 9.1e-009
>pbeta(0,9,2) for p=1
[1] 0
>49/50
[1] 0.98
>0.01*0.98*pbeta(0.95,9,2)
[1] 0.009758191 for p=0.01
>0.05*0.98*pbeta(0.95,9,2)
[1] 0.04477922 for p=0.05
>0.10&0.98*pbeta(0.90,9,2)
[1] 0.0721377 for p=>0.10
>0.15*98*pbeta(0.85,9,2)
[1] 0.08001207 for p=0.15
>0.20*0.98*pbeta(0.80,9,2)
[1] 0.07365869 for p=0.20
>0.25*0.98*pbeta(0.75,9,2)
[1] 0.05978618 for p=0.25
>0.30*0.98*pbeta(0.70,9,2)
[1] 0.04389665 for p=0.30
>0.40*0.98*pbeta(0.60,9,2)
[1] 0.0181721 for p=0.40
>0.50*0.98*pbeta(0.50,9,2)
[1] 0.005263672 for p=0.50
>0.60*0.98*pbeta(0.40,9,2)
[1] 0.0009865003 for p=0.60
>0.70*0.98*pbeta(30,9,2)
[1] 0.00009856853 for p=0.70
>0.80*0.98*pbeta(0.20,9,2)
[1] 3.291546e-006 for p=0.8
>0.90*0.98*pbeta(0.10,9,2)
[1] 8.0262e-009 for p=0.90
>1.00*0.98*pbeta(0,9,2).
[1] 0 for p=1.00
```

On the basis of the above data a graphical representation of AOQ curve for $N = 500$, $n = 10$, $c = 1$ is given below. The ordinate of the curve represents the average outgoing quality (AOQ).

The method of AOQ curve is considerably simplified under Poisson approximation of Binomial process curve. In this case, suppose p, the proportion of defective, is small and the mean number of defective is np, then from (1.7.2), (1.4.1) and Poisson approximation to binomial, we have

$$(\text{AOQ})_{\text{rep}} = \frac{p(N-n)}{N}\sum_{r=0}^{c}\frac{e^{-np}.(np)^r}{r!} = p\frac{(N-n)}{N}\frac{1}{\Gamma(c+1)}\int_{np}^{\infty} e^{-z}z^c dz.$$

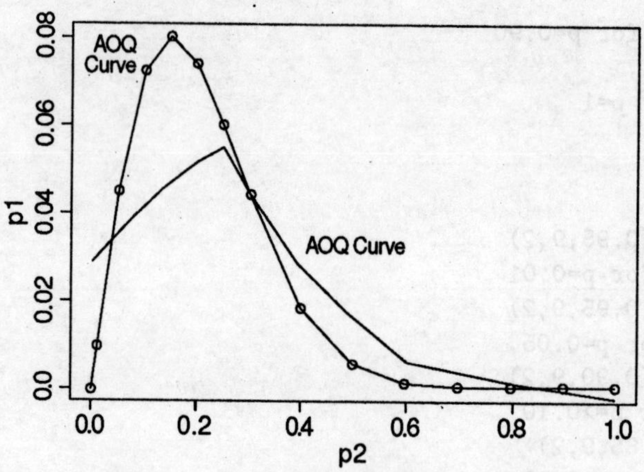

Fig 4

Putting

$$z = \frac{\chi^2}{2} \Rightarrow (\text{AOQ})_{\text{rep}} = \frac{p(N-n)}{N\Gamma(c+1)}\int_{2np}^{\infty} e^{-\frac{\chi^2}{2}}.(\chi^2/2)^{\frac{2(c+1)}{2}-1}\, d(\chi^2/2)$$

$$= p\frac{(N-n)}{N} P(\chi^2 > 2np \mid \nu = 2(c+1)), \qquad (1.8.3)$$

where ν is the degree of freedom.

Given c, N and n, one can plot the curve for different p which gives AOQ.

1.9 Average Outgoing Quality Limits (AOQL)

This refers to the maxima of AOQ curve.

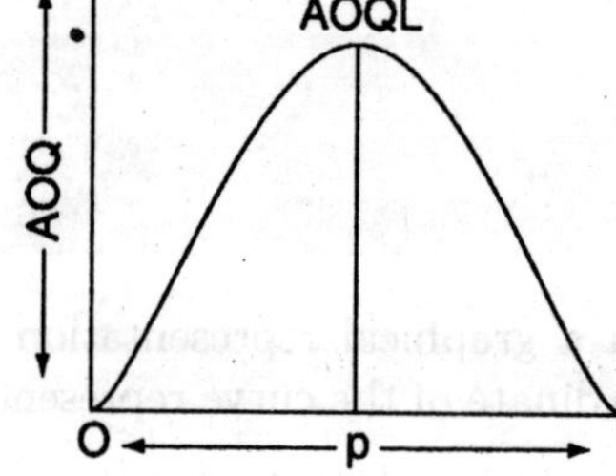

Fig 5

Given $(\text{AOQ})_{\text{rep}} = N^{-1}[p(N-n)]P(p)]$ from (1.7.3), where $P(p)$ represents the probability of acceptance of the lot, AOQL is given by

$$\frac{d}{dp}\left(\frac{p(N-n)P(p)}{N}\right)=0 \Rightarrow \frac{1}{N}\left[(N-n)P(p)+p(N-n)\frac{dP(p)}{dp}\right]=0$$

$$\Rightarrow p\frac{dP(p)}{dp}=-P(p)$$

$$\Rightarrow p=\frac{-P(p)}{dP(p)/dp}=\frac{-1}{d(\log P(p))/dp}. \tag{1.9.1}$$

▶▶ NOTE

AOQL corresponds to P(p) which is maximum but dP(p)/dp is minimum. The value of p satisfying (1.9.1) corresponds to AOQL.

Example 2 *Obtain* AOQL *if the number of the defective conforms to a Poisson distribution with parameter np.*

Solution: Here

$$(\text{AOQ})_{\text{rep}} = \frac{p(N-n)}{N}\sum_{r=0}^{c}\frac{e^{-np}(np)^r}{r!} \quad [\text{from } (1.4.1)]$$

$$= p\frac{(N-n)}{N}\frac{1}{\Gamma(c+1)}\int_{np}^{\infty}e^{-z}z^c dz.$$

Hence

$$\frac{d(\text{AOQ})}{dp}=\frac{(N-n)}{N\Gamma(c+1)}\int_{np}^{\infty}e^{-z}z^c dz+\frac{p(N-n)}{N\Gamma(c+1)}[-ne^{-np}(np)^c]=0$$

$$\Rightarrow \frac{(N-n)}{N}P\left(\chi^2>2np \mid \nu=2(c+1)\right)+\frac{p(N-n)}{N\Gamma(c+1)}\left[-ne^{-np}(np)^c\right]=0.$$

Given n, c and N, the above equation may be solved by successive iteration which corresponds to AOQL.

1.10 Goodness of a Sampling Inspection Plan

As discussed earlier, the goodness of a plan is judged from three basic criteria, viz.,

(*i*) OC curve of the plan.

(*ii*) ASN which gives an idea about the cost of the plan.

(*iii*) AOQ which gives information into the quality of the lots selected.

However, the criterion (*iii*) is dependent on (*i*). An alternative to OC function is known as *Average Run Length* (ARL).

ARL is defined as the average number of non-defective lots in succession till a defective lot is arrived. The successive number of non-defective lots is a measure of run length. The average run length taken over all the lots, i.e., ARL is taken as an alternate index to OC function.

If $P(p)$ represents the probability of acceptance of a lot with proportion of defective p, then run length r of a lot (a random variable) and its probability distribution is given by

$$[P(p)]^{r-1}[1-P(p)], \tag{1.10.1}$$

where r is the successive number of non-defective; $r = 0, 1, 2, 3, \ldots$

$$\begin{aligned} E(r) = \text{ARL} &= \sum_{r=0}^{\infty} r[P(p)]^{r-1}[1-P(p)] \\ &= [1-P(p)] + 2[P(p)][1-P(p)] + 3[P(p)^2][1-P(p)] + \ldots \\ &= [1-P(p)][1+2P(p)+3[P(p)]^2+4[P(p)]^3+\ldots] \\ &= [1-P(p)][1-P(p)]^{-2} \\ &= [1-P(p)]^{-1}. \end{aligned} \tag{1.10.2}$$

Of course, in this case the OC function given by $P(p)$ and the ARL given by $[1-P(p)]^{-1}$ serve the same purpose and ARL is a direct function of OC function. In general, ARL serves more information than OC function.

Sometimes, it is not necessary to test all the units but still we may be able to take decision before inspecting the entire sample. Details are given as follows.

1.11 Curtailed and Semi-curtailed Sampling Inspection Plan

There are two cases of curtailed sampling plan, viz., curtailed sampling plan with (*i*) rejection of the lot and (*ii*) acceptance of the lot.

1.11.1 Curtailed sampling plan with rejection of the lot

Consider a situation, where $(i-1)$ units from a sample of size n have been inspected and found that c (= acceptance number) of them are defective. If the ith unit also turns out to be defective, we have $(c+1)$ units are defective units in a sample of size $i < n$. Since the number of defective exceeds c, one can reject the lot at this step without proceeding with further inspection. This is a case of curtailment of the sample size in the sampling inspection plan with the decision of rejecting the lot.

1.11.2 Curtailed sampling plan with acceptance of the lot

Consider another situation, where we curtail the sample size while taking a decision of accepting the lot. Suppose, we have inspected $(n-c+j-1)$ units from a sample of size n and found j of them are defective. Also it is found that the next unit, i.e., $(n-c+j)$th unit is non-defective. Thus we have a total of j defective units in a sample inspection of $(n-c+j)$ units. The residual number of non-inspected units are $n-(n-c+j) = (c-j)$. Now even if all the $(c-j)$ units are defective, the total number of defective units in a sample of size n can be at most $(j+c-j) = c$. This number while not exceeding the Acceptance Number (AN) c, the lot may be accepted on the basis of a sample inspection of size $(n-c+j)$, $(j \leq c)$. This is a case of curtailed sampling inspection plan with acceptance. The Average Sample Number (ASN) of a single sampling plan is thus given by

$$\underbrace{\sum_{i=c+1}^{n} i\binom{i-1}{c} p^{c+1}(1-p)^{i-1-c}}_{\text{curtailment with rejection}} + \underbrace{\sum_{j=0}^{c}(n-c+j)\binom{n-c+j-1}{j} p^{j}(1-p)^{n-c}}_{\text{curtailment with acceptance}} \tag{1.11.1}$$

1.11.3 Semi-curtailed sampling plan

The semi-curtailed sampling plan is based on the principle that we may reject the lot without inspecting all the items but accept only by inspecting each and every item in the sample. The former refers to curtailment and the latter refers to entire sample (i.e., without curtailment). This is precisely the semi-curtailed sampling inspection plan and ASN of the same with the same reasoning as in the case of curtailment is given by

$$\underbrace{\sum_{i=c+1}^{n} i\binom{i-1}{c} p^{c+1}(1-p)^{i-1-c}}_{\text{rejection with curtailment}} + \underbrace{\sum_{r=0}^{c} r\binom{n}{r} p^{r}(1-p)^{n-r}}_{\text{acceptance with curtailment}} \tag{1.11.2}$$

1.11.4 Method of determining a sampling inspection plan (n, c)

We have discussed in Example 1 the basic problem of finding the plan, i.e., (n, c) by assuming a large sample approximation of Binomial distribution to Normal distribution. However, in this section, three exact methods are described. First one deals with binomial sampling plan which becomes easily operable when a decision about n has already been taken. The second method deals when a Poisson approximation to binomial is assumed. Finally, the third method, a Bayesian approach to the problem, is shown under a prior distribution of p based on the equality of posterior losses in rejection and acceptance.

First method: Binomial sampling plan

Consumer's risk in a binomial sampling inspection plan is given by

$$\sum_{r=0}^{c}\binom{n}{r} p^{r}(1-p)^{n-r} = \beta.$$

Using incomplete Beta function, the above can be expressed as

$$\frac{1}{B(n-c, c+1)}\int_0^{1-p_2} t^{n-c-1}(1-t)^c\, dt = \beta,$$

where p_2 is the consumer value of p.

We denote the above as p beta $(1-p_2, n-c, c+1) = \beta$ ($S-$ plus notation).

Similarly, producer's risk α is given by

$$1-\frac{1}{B(n-c,c+1)}\int_0^{1-p_1} t^{n-c-1}(1-t)^c\,dt=\alpha,$$

where p_1 is the producer's value of p

$$\Rightarrow \quad \frac{1}{B(n-c,c+1)}\int_0^{1-p_1} t^{n-c-1}(1-t)^c\,dt=1-\alpha$$

$$\Rightarrow \quad p\text{ beta }(1-p_1,n-c,c+1)=1-\alpha.$$

$$\therefore \frac{p\text{ beta }(1-p_1,n-c,c+1)}{p\text{ beta }(1-p_2,n-c,c+1)}=\frac{1-\alpha}{\beta}. \tag{1.11.3}$$

Given α, β, p_1 and p_2, it is possible to obtain c (when n is known) by $S-$ plus or any other program. A numerical illustration for assumed value of $c(=5)$ is given below:

Numerical illustration

Suppose, producer's value of $p=p_1=0.01$ and the consumer's value $p=p_2=0.04$ and both $\alpha=\beta=0.05$, then our subject is to find out the sampling plan (n,c). We have,

$$(1-\beta)/\alpha=19.$$

Then $\text{pbeta}(0.96,n-c,c+1)=0.05$ and $\text{pbeta}(0.99,n-c,c+1)=0.95$.

That is, $\text{pbeta}(0.99,n-c,c+1)/\text{pbeta}(0.96,n-c,c+1)=19$.

To obtain n and c by $S-$plus computation is shown as below: We try with a trial value of $n=262$ and $c=5$. Taking $n=262,261,260\ldots$ and $c=5\Rightarrow n-c=257,256,255,\ldots$ and $c+1=6$, we have $\text{pbeta}(0.99,257,6)=0.9503735$, $\text{pbeta}(0.96,257,6)=0.04772972$.

```
We have >0.9503735/0.04772972
[1] 19.91157
>pbeta (0.99,256,6)
[1] 0.9511146
>pbeta (0.96,256,6)
[1] 0.04888074
>0.9511146/0.04888074
[1] 19.45786
>pbeta. (0.99,255,6)
[1] 0.9518489
>pbeta (0.95,255,6)
[1] 0.05005674
>0.9518489/0.05005674
[1] 19.0154
>pbeta (0.99,254,6)
[1] 0.9525763
>pbeta (0.96,254,6)
[1] 0.05125819
>0.9525763/0.05125819
[1] 18.58388
```

From this observations, it follows that $\text{pbeta}(0.99,255,6)/\text{pbeta}(0.96,255,6)=19.0154$ which is closest to $(1-\beta)/\alpha=19$.

Hence, $n-c=255$, $c+1=6\Rightarrow n=260$ and $c=5$. Thus $(n,c)=(260,5)$.

Remarks

The successive approximation process for obtaining a plan (n, c) can be much simplified under Poisson approximation to Binomial.

Second method: Sampling plan based on Poisson approximation to binomial

For small p_1 and p_2 and large n, subject to the condition that the respective means are fixed at np_1 and np_2 respectively (i.e., justifying Poisson approximation to binomial), we have,

$$\sum_{r=0}^{c} \frac{e^{np_1}\,(np_1)^r}{r!} = 1-\alpha \tag{1.11.4}$$

$$\text{and} \sum_{r=0}^{c} \frac{e^{-np_2}\,(np_2)^r}{r!} = \beta. \tag{1.11.5}$$

Equations (1.11.4) and (1.11.5) are equivalently written as

$$\frac{1}{\Gamma(c+1)}\int_{np_1}^{\infty} e^{-z}z^c dz = 1-\alpha. \tag{1.11.6}$$

$$\frac{1}{\Gamma(c+1)}\int_{np_2}^{\infty} e^{-z}z^c dz = \beta. \tag{1.11.7}$$

Again, equations (1.11.6) and (1.11.7) can be expressed as

$$P[\chi^2 > 2np_1 \mid \nu = 2(c+1)] = 1-\alpha \tag{1.11.8}$$

$$\text{and } P[\chi^2 > 2np_2 \mid \nu = 2(c+1)] = \beta. \tag{1.11.9}$$

Equations (1.11.9), (1.11.8) gives the ratio of the critical values of χ^2—covering $100(1-\beta)\%$ area to the critical value of χ^2 covering $100\alpha\%$ of the area which is equal to the ratio of $2np_2$ to $2np_1$. Symbolically,

$$\frac{\chi^2(1-\beta)}{\chi^2\alpha} = \frac{2np_2}{2p_1} = \frac{p_2}{p_1} \text{ for } \nu = 2(c+1) \text{ degrees of freedom.}$$

Thus the value of c is obtainable from χ^2-table. Finally, the value of n can be got either by using (1.11.8) or (1.11.9) which may differ but not very widely. A numerical illustration with the same data as given in the earlier numerical illustration (which assumed only binomial sampling inspection plan without its Poisson approximation) is given below.

Numerical illustration

We have $p_1 = 0.01$, $p_2 = 0.04$, $\alpha = 0.05$ and $\beta = 0.05$. In this case, 95% value of χ^2 to 5% value of χ^2 for $\nu = 2(c+1)$, degrees of freedom is in the ratio of 4 : 1; χ^2-table shows that the ratio of two χ^2 at 95% and 5% levels are in the ratio 4 : 1 under degrees of freedom $\nu = 2(c+1) = 12$. This gives $c = 5$. Again (1.11.8) shows $2np_1 = 5\%$ critical values of χ^2 with 5 degrees of freedom $= 5.23$ (from χ^2-table) $\Rightarrow 2n(0.01) = 5.23 \Rightarrow n = 5.23/0.02 = 261.5$.

Similarly, (1.11.9) shows $2np_2 - 95\%$ critical value of χ^2 with 5 degrees of freedom $= 21.03$ (from χ^2-table) $\Rightarrow 2n(0.04) = 21.03 \Rightarrow n = 21.03/0.08 = 262.5$. Thus we have $n = 262$ approximately from either (1.11.8) or (1.11.9).

Remarks

The two results, viz., $(n, c) = (260, 5)$ for binomial sampling inspection plan and $(n, c) = (262, 5)$ for Poisson approximation of the plan compare favourably.

Third Method: Bayesian approach to the sampling inspection plan

The Bayesian approach in obtaining a sampling plan (n, c) comprises of obtaining posterior distribution of p given r corresponding to a specific sampling inspection plan (ex. binomial, Poisson, negative binomial, etc.); and then by equating posterior expected losses in acceptance and rejection of the lot.

Notation:

$N \equiv$ lot size
$n \equiv$ sample size
$p \equiv$ proportion of the defective
$f \equiv$ loss per item of the defective
$g \equiv$ gain per good item

Then the total gain/loss is given by

$$\pi = [N(1-p)g - Npf] \tag{1.11.10}$$

$$= Ng - p(Ng + Nf).$$

We put $Ng = kp_0$ and $Ng + Nf = k$, then $\pi = k(p_0 - p)$.
Hence

$$p_0 = \frac{Ng}{k} = \frac{Ng}{Ng + Nf} = \frac{g}{g+f}. \tag{1.11.11}$$

Next, we have the loss depending on the unknown situations as well as the decision taken as in the following table.

Table 1: Amount of loss incurred

	Situation	
Decision	$p < p_0$	$p > p_0$
Accept	0	$k \mid p_0 - p \mid$
Reject	$k \mid p - p_0 \mid$	0

We have the sampling distribution of the number of the defective r, given r being binomial with parameters (n, p).

$$P(r \mid p) = {}^nC_r(1-p)^{n-r}p^r$$

and the best prior distribution of p under binomial sampling

$$f(p) = \frac{p^{\alpha-1}(1-p)^{\beta-1}}{B(\alpha, \beta)}. \tag{1.11.12}$$

Therefore, the posterior distribution of p, r being given, is

$$\begin{aligned}\phi(p \mid r) &= \frac{P(r \mid p) f(p)}{\int_0^1 P(r \mid p) f(p)\, dp} \\ &= \frac{p^{\alpha+r-1}\,(1-p)^{\beta+n-r-1}/B(\alpha,\beta)}{\int_0^1 p^{\alpha+r-1}\,(1-p)^{\beta+n-r-1}\,dp/B(\alpha,\beta)} \\ &= \frac{p^{\alpha+r-1}\,(1-p)^{n-r+\beta-1}}{B(\alpha+r, n-r+\beta)}.\end{aligned}$$

Posterior expected loss in acceptance is

$$\frac{k\int_{p_0}^1 \mid p-p_0 \mid p^{\alpha+r-1}\,(1-p)^{n-r+\beta-1}\,dp}{B(\alpha+r, n-r+\beta)}. \tag{1.11.13}$$

Posterior expected loss in rejection is

$$\frac{k\int_0^{p_0} \mid p_0-p \mid p^{\alpha+r-1}\,(1-p)^{n-r+\beta-1}\,dp}{B(\alpha+r, n-r+\beta)}. \tag{1.11.14}$$

The value of $r = c$ is decided by equating (1.11.13) and (1.11.14),

$$\begin{aligned}&\Rightarrow \int_0^{p_0} (p_0-p)p^{\alpha+r-1}\,(1-p)^{n-r+\beta-1}\,dp = \int_{p_0}^1 (p-p_0)p^{\alpha+r-1}\,(1-p)^{n-r+\beta-1}\,dp \\ &\Rightarrow \int_0^{p_0} (p_0-p)p^{\alpha+r-1}\,(1-p)^{n-r+\beta-1}\,dp + \int_{p_0}^1 (p_0-p)p^{\alpha+r-1}\,(1-p)^{n-r+\beta-1}\,dp = 0 \\ &\Rightarrow \int_0^1 (p_0-p)p^{\alpha+r-1}\,(1-p)^{n-r+\beta-1}\,dp = 0 \\ &\Rightarrow p_0\int_0^1 p^{\alpha+r-1}\,(1-p)^{n-r+\beta-1}\,dp = \int_0^1 p^{\alpha+r}\,(1-p)^{n-r+\beta-1}\,dp \\ &\Rightarrow p_0[B(\alpha+r, n-r+\beta)] = B(\alpha+r+1, n-r+\beta) \\ &\Rightarrow p_0 = \frac{B(\alpha+r+1, n-r+\beta)}{B(\alpha+r, n-r+\beta)} \\ &\quad = \frac{\Gamma(\alpha+r+1)\Gamma(n-r+\beta)}{\Gamma(\alpha+r)\Gamma(n-r+\beta)} \times \frac{\Gamma(\alpha+n+\beta)}{\Gamma(\alpha+1+n+\beta)} = \frac{\alpha+r}{\alpha+n+\beta} \\ &\Rightarrow p_0(\alpha+n+\beta) = (\alpha+r) \\ &\Rightarrow p_0(\alpha+\beta)+np_0-\alpha = r = c.\end{aligned}$$

Now

$$E(p) = \frac{\alpha}{\alpha+\beta} \text{ [from (1.11.12)].}$$

Replacing p_0 by $E(p)$, we have

$$c \equiv \frac{\alpha}{\alpha+\beta}\,(\alpha+\beta)+np_0-\alpha, \text{ i.e., } c \equiv np_0.$$

Again

$$p_0 = \frac{g}{g+f} \Rightarrow c \equiv \frac{ng}{g+f}.$$

Now, to obtain the specific value of c, we are required to find n. We have, total loss = cost of sampling + cost of wrong decision.

Let the cost of inspection per unit be unity, then

$$\begin{aligned}
\pi &= n + \int_{p_0}^{1} k(p-p_0)P(p)f(p)\,dp + \int_0^{P_0} k(p_0-p)[1-P(p)]f(p)\,dp, \\
&\quad \text{where } f(p) = P(r \mid p) \text{ ad } f(p) \text{ is the the prior density of } p \\
&= n + \int_{p_0}^{1} k(p-p_0)P(p)f(p)\,dp + \int_0^{p_0} k(p_0-p)f(p)\,dp - \int_0^{p_0} k(p_0-p)P(p)f(p)\,dp \\
&= n + \int_0^{p_0} k(p-p_0)P(p)f(p)\,dp + \int_{p_0}^{1} k(p-p_0)P(p)f(p)\,dp - \int_0^{p_0} k(p-p_0)f(p)\,dp \\
&= n + \int_0^{1} k(p-p_0)P(p)f(p)\,dp - \int_0^{p_0} k(p-p_0)f(p)\,dp.
\end{aligned}$$

To minimize π with respect to n we set $\frac{d\pi}{dn} = 0$, which gives

$$\frac{d\pi}{dn} = 0 = 1 + \int_0^1 k(p-p_0)\frac{dP(p)}{dn}f(p)\,dp. \tag{1.11.15}$$

One can solve (1.11.15) by choosing a particular form of $P(p)$ and $f(p)$. We give a specific example to find n.

Example 3 *Let* $P(r \mid p) = \sum_{r=0}^{c}\binom{n}{r}p^r(1-p)^{n-r}$ *and* $f(p) = 1$ *in* $(0,1)$, *the distribution of* p *being uniform; given that* n *is large and* $p \to 0$ *subject to fixed* np. *Obtain* n *and* c *of the sampling inspection plan by Bayesian approach.*

Solution: In this case, for large n and $p \to 0$ the binomial distribution converges to Poisson

$$\frac{d\pi}{dn} = 0 \Rightarrow 1 + \int_0^1 k(p-p_0)\frac{d}{dn}\sum_{r=0}^{c} e^{-np}\frac{(np)^r}{r!}\,dp = 0.$$

We have the estimating equation for n as

$$1+\int_0^1 k(p-p_0)\left(\frac{d}{dn}\sum_{r=0}^{c} e^{-np}\frac{(np)^r}{r!}\right)dp = 0$$

$$\Rightarrow\ 1+\int_0^1 k(p-p_0)\left(\frac{d}{dn}\frac{1}{\Gamma(c+1)}\int_{np}^{\infty} e^{-z}z^c dz\right)dp = 0$$

$$\Rightarrow\ 1+\frac{k}{\Gamma(c+1)}\int_0^I (p-p_0)(-p)e^{-np}(np)^c dp = 0$$

$$\Rightarrow\ 1+\frac{k}{\Gamma(c+1)}\int_0^n\left(-\frac{m}{n}\right)\left(\frac{m}{n}-p_0\right)e^{-m}m^c d\left(\frac{m}{n}\right) = 0$$

$$\Rightarrow\ 1-\frac{k}{\Gamma(c+1)n^3}\int_0^n m(m-np_0)e^{-m}m^c dm = 0$$

$$\Rightarrow\ 1 = \frac{k}{\Gamma(c+1)n^3}\left(\int_0^n e^{-m}m^{c+2}\,dm - np_0\int_0^n e^{-m}m^{c+1}\,dm\right)$$

$$1 \equiv \frac{k}{\Gamma(c+1)n^3}[\Gamma(c+3)-np_0\Gamma(c+2)], \text{ for large } n$$

$$1 = \frac{k(c+2)(c+1)}{n^3} - \frac{kp_0(c+1)}{n^2}$$

$$\Rightarrow\ n^3 = k(c+1)[c+2-np_0].$$

Putting $np_0 = c$ we have,

$$n^3 = 2k(c+1) \Rightarrow \frac{n^3}{2k} = c+1 = np_0+1, \tag{1.11.16}$$

where $k = N(g+f)$ and $p_0 = \frac{g}{g+f}$ are known quantities.

By solving n one can get $c = np_0$.

1.12 Some Observations on the Sampling Inspection Plan

A question that basically arises is that what is the efficacy of the sampling inspection plan? Does the quality of the plan depend on the choice of the process curve (i.e., the long term distribution of quality of the lot in terms of the number of the defective)?

To investigate the same, let us take a hypothetical case when the lot size is N with R number of the defective. Then the probability that a sample of size n from the lot will have r number of the defective, is

$$\frac{\binom{n}{r}\binom{N-n}{R-r}}{\binom{N}{R}}.$$

Suppose, the process curve is binomial, i.e., the probability of R number of the defective in the lot of size N is given by

$$\binom{N}{R} p^R(1-p)^{N-R}.$$

Then the joint probability of obtaining (N, R) and (n, r) is given by

$$\frac{\binom{n}{r}\binom{N-n}{R-r}}{\binom{N}{R}}\binom{N}{R}p^R q^{N-R} = \left[\binom{n}{r}p^r q^{n-r}\right]\left[\binom{N-n}{R-r}p^{R-r}q^{N-n)-(R-r)}\right], \tag{1.12.1}$$

which implies that the random variables r and $(R-r)$ are independently distributed and therefore, uncorrelated. More precisely, it shows that if the process curve is binomial, the number of the defective in the sample does not give an idea of the residual number of the defective in the parent universe, i.e., $(R-r)$. However, in case of Leptokurtic and Platykurtic process curves, the same correlations may be shown to be negative and positive respectively (Mood, 1943). A natural conclusion then comes that under constant level of quality of production, the traditional sampling inspection plan is not effective. However, the plan has efficacy if the quality varies from lot to lot.

Another question that is worthwhile to consider is that what is the probability that a lot of quality, *say*, $p = p_0$ or $p \geq p_0$, is present in the lot already accepted?

For this, we require the prior information of p or the process curve for p and the OC function, given r. Let us denote them by $R(p)$ and $P(p \mid r)$ respectively (r = Number of the defective). In fact, the probability distribution of quality in the accepted lots

$$= \frac{P(p \mid r)R(p)}{\int_0^1 P(p \mid r)R(p)\,dp}. \tag{1.12.2}$$

The probability of a quality $p \geq p_0$ in accepted lots is

$$= \frac{\int_{p_0}^1 P(p \mid r)R(p)\,dp}{\int_0^1 P(p \mid r)R(p)\,dp}. \tag{1.12.3}$$

Thus in deciding the level of quality in the accepted lots, OC function alone is not sufficient, but we require a knowledge of the process curve. The Average Outgoing Quality (AOQ) of the acceptance sampling plan when quality varies from lot to lot conforming to a distribution $R(p)$ is given by

$$\text{AOQ} = \frac{\int_0^1 pP(p \mid r)R(p)\,dp}{\int_0^1 p(p \mid r)R(p)\,dp} = E(p). \tag{1.12.4}$$

The variance of the quality of the accepted lots

$$= \frac{\int_0^1 [p - E(p)]^2 P(p \mid r)R(p)\,dp}{\int_0^1 P(p \mid r)R(p)\,dp} \tag{1.12.5}$$

1.13 Inspection by Variables of Single Sampling Plan

Hitherto, we have been examining the efficacy of sampling inspection plan by attributes (more specifically by the proportion of the defective). We now consider the sampling inspection plan of items for which the standard is defined in terms of measurements (*say*, length, weight, etc.). The standard demanded for acceptance is given by the specification limit. We may have single or double specification limits. Let us illustrate these concepts by the following example.

Suppose, while manufacturing steel balls of specified diameter, *say*, 0.2 cm on the average, it is told that the balls of diameter less than 0.19 cm will not be accepted. Then it is a problem of single specification. On the other hand, if it is told that only the balls with diameter lying between 0.19 cm and 0.21 cm will be accepted, then we have a double specification limit. We first consider sampling inspection plan by variables under single specification limit.

1.13.1 Single specification limit

Let L be a limit below which products are unacceptable or considered defective. Here we require $P(X < L)$ given $X \sim N(\mu, \sigma^2)$ and σ^2 is known.

We have

$$P\left(\frac{X-\mu}{\sigma} < \frac{L-\mu}{\sigma}\right), \text{ if } X < L. \tag{1.13.1}$$

We denote

$$P\left(\frac{X-\mu}{\sigma} < \frac{L-\mu}{\sigma}\right) = \Phi\left(\frac{L-\mu}{\sigma}\right).$$

Also

$$\Phi\left(\frac{L-\mu}{\sigma}\right) = 1 - \Phi\left(\frac{\mu-L}{\sigma}\right). \tag{1.13.2}$$

We test $H_0: \ \theta = \theta_1$ against $H_1: \ \theta = \theta_2$; where θ is the true proportion of the defectives. We can write

$$\tilde{p} = 1 - \Phi\left(\frac{\bar{X}-L}{\sigma}\right) \tag{1.13.3}$$

as an estimate of the proportion of the defective. A better estimate obtained by Leibermann and Resnikoff (1955) is given by

$$\tilde{p} = 1 - \Phi\left(\frac{\bar{X}-L}{\sigma}\sqrt{\frac{n}{n-1}}\right). \tag{1.13.4}$$

If $\tilde{p} < p*$ ($0 \le p^* \le 1$; where p^* is a fixed proportion already decided by the producer and consumer for the acceptance of the lot), i.e., if

$$\tilde{p} = 1 - \Phi\left(\frac{\bar{X}-L}{\sigma}\sqrt{\frac{n}{n-1}}\right) < p^* \tag{1.13.5}$$

the lot is accepted; otherwise it is rejected.

$$\Rightarrow \tilde{p} = 1 - \Phi\left(\frac{\bar{X}-L}{\sigma}\sqrt{\frac{n}{n-1}}\right) < \int_{-\infty}^{-Zp^*} f dx = \Phi(-Z_{p^*}) = 1 - \Phi(z_{p^*})$$

f being the probability density function of standardized normal distribution and $(-z_{p^*})$ is the abscissa point for which $(-\infty, -z_{p^*})$ corresponds to a probability p^*.

$$\Rightarrow \frac{\bar{X}-L}{\sigma}\sqrt{\frac{n}{n-1}} > z_{p^*}, \text{ where } p^* = \frac{1}{\sqrt{2\pi}}\int_{z_{p^*}}^{\infty} e^{-\tau^2/2}\, d\tau$$

$$\Rightarrow P\left[\bar{X} > L + z_{p^*}\cdot\sigma\sqrt{\frac{n-1}{n}}\right] \text{ is the OC function.} \tag{1.13.6}$$

$$\therefore P\left(\bar{X} > L + z_{p^*}\cdot\sigma\sqrt{\frac{n-1}{n}}\right) = \begin{cases} 1-\alpha & \text{when } H_0 \text{ is true,} \\ \beta & \text{when } H_1 \text{ is true.} \end{cases} \tag{1.13.7}$$

Next to find n and p^*, we assume θ_1 proportion of the defective is accepted with probability $(1-\alpha)$, and θ_2 proportion with probability β. In other words, $(1-\alpha)$ proportion of lots with the defective proportion θ_1, and β proportion of lots with the defective proportion θ_2 are accepted under the OC function.

Now for a lot to have θ_1 proportion of the defective when σ is known and the mean $\mu = \mu_1$, is given by,

$$\Phi\left(\frac{L-\mu_1}{\sigma}\right) = \theta_1 \Rightarrow \left(\frac{L-\mu_1}{\sigma}\right) = -z_{\theta_1}. \tag{1.13.8}$$

Similarly,

$$\Phi\left(\frac{L-\mu_2}{\sigma}\right) = \theta_2,$$

i.e., to have proportion of the defective θ_2, the mean will be μ_2.

$$\Rightarrow \left(\frac{L-\mu_2}{\sigma}\right) = -z_{\theta_2}. \tag{1.13.9}$$

Now

$$P\left(\bar{X} > L + z_{p^*}\cdot\sigma\sqrt{\frac{n-1}{n}}\right) = 1-\alpha, \text{ when } \mu = \mu_1$$

$$\Rightarrow \quad P\left(\frac{\bar{X}-\mu_1}{\sigma/\sqrt{n}} > \frac{L-\mu_1}{\sigma/\sqrt{n}} + z_{p^*}\sqrt{n-1}\right) = 1-\alpha$$

$$\Rightarrow \quad P\left(\frac{\bar{X}-\mu_1}{\sigma/\sqrt{n}} > -z_{\theta_1}\sqrt{n} + z_{p^*}\sqrt{n-1}\right) = 1-\alpha \text{ [from (1.13.8)]}$$

$$\Rightarrow \quad -z_{\theta_1}\sqrt{n} + z_{p^*}\sqrt{n-1} = -z_\alpha \tag{1.13.10}$$

$$\text{and} \quad P\left(\frac{\bar{X}-\mu_2}{\sigma/\sqrt{n}} > \frac{L-\mu_2}{\sigma/\sqrt{n}} + z_{p^*}\sqrt{n-1}\right) = \beta$$

$$\Rightarrow -z_{\theta_2}\sqrt{n} + z_{p^*}\sqrt{n-1} = z_\beta \text{ [from (1.13.9)]} \tag{1.13.11}$$

(1.13.10) and (1.13.11) $\Rightarrow$

$$\Rightarrow \quad -z_{\theta_1}\sqrt{n} + z_{\theta_2}\sqrt{n} = -z_\alpha - z_\beta$$

$$\Rightarrow \quad -\sqrt{n}\,(z_{\theta_1} - z_{\theta_2}) = -(z_\alpha + z_\beta)$$

$$\Rightarrow \quad \sqrt{n} = \frac{(z_\alpha + z_\beta)}{(z_{\theta_1} - z_{\theta_2})}$$

$$\Rightarrow n = \frac{(z_\alpha + z_\beta)^2}{(z_{\theta_1} - z_{\theta_2})^2}. \tag{1.13.12}$$

Also,

$$-z_{\theta_1} + z_{p^*}\sqrt{\frac{n-1}{n}} = -\frac{z_\alpha}{\sqrt{n}} \text{ [from (1.13.10)]}$$

$$\Rightarrow \quad z_{p^*}\sqrt{\frac{n-1}{n}} = z_{\theta_1} - \frac{z_\alpha}{\sqrt{n}}. \tag{1.13.13}$$

Similarly,

$$\Rightarrow \quad z_{p^*}\sqrt{\frac{n-1}{n}} = z_{\theta_2} + z_{\beta/\sqrt{n}} \quad \text{[from (1.13.11)]} \tag{1.13.14}$$

It may be noted that the solution of p^* will not be the same based on (1.13.13) and (1.13.14). However, the mean value of p^* based on both the equations will be satisfactory. The OC function can also be derived as follows:

If there are 100 per cent defective in the lot, the mean must be at $\mu = L + z_p\sigma$.

The probability of acceptance is the probability for which (1.13.2) is true, i.e.,

$$P\left(\bar{X} > L + z_{p^*}\sigma\sqrt{\frac{n-1}{n}}\middle|\bar{X} \sim N(\mu, \sigma^2/n)\right) = \Phi\left(\sqrt{n}\, z_p - \sqrt{n-1}\, z_{p^*}\right), \tag{1.13.15}$$

which is the OC function of the plan.

1.13.2 Double specification limit

Let us consider a two-sided specification given by (L, U). If $(U - L)/\sigma$ is sufficiently large so that a sampling plan can be constructed by combining the two one-sided tests as presented in the preceding section and all the formulas will remain approximately correct. However, there exists a minimum percentage of the defective corresponding to the mean in the centre of the range (L, U). The percentage of the defective is then

$$1 - \Phi\left(\frac{L-U}{\sigma}\right) \cong 2\Phi\left\{\left(\frac{L-U}{2\sigma}\right)\right\}.$$

Now if $(U - L)/\sigma$ is reasonably small one can reject the lot outright. Again for large values of $(U - L)/\sigma$, one can accept the lot. But for intermediate values of $(U - L)/\sigma$ there arises some difficultly. It is necessary to have a kind of relationship between total percentage of the defective and the percentage of the defective at each tail for taking decision on such intermediate values.

Without loss of generality, let us put $L = 0$ and $U = 1$ so that σ is represented in the same scale. Since $U - L$ is held fixed, the value of σ with which we are interested may not lead to very high or low values of $(U - L)/\sigma$. Let the lot have a mean X, then the proportion of the defective is

$$p = \Phi(-X/\sigma) + \Phi[(X-1)/\sigma]. \tag{1.13.16}$$

The two terms on the R.H.S. of the above equation represent the proportion of the defective at the two ends. Equation (1.13.16) when graphically plotted gives a curve which is symmetrical about $(L + U)/2$. This is shown in Fig 6.

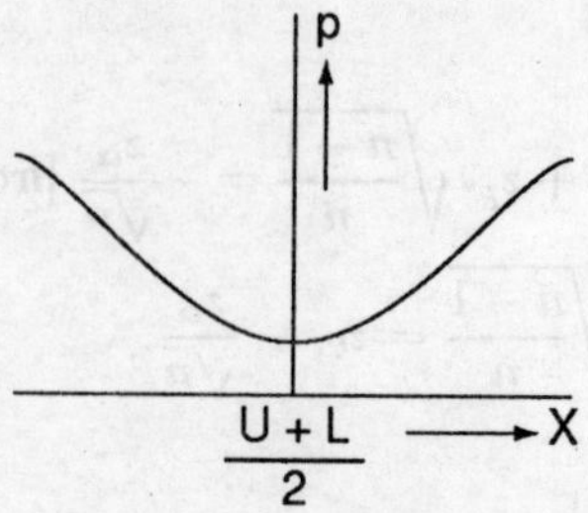

Fig 6 Double Specification Limits of the Proportion of the Defectives

Using the above graph one can obtain the proportion of the defective at either end for a given value of p.

In the next place, suppose, our plan comprises of accepting a proportion of $(1-\alpha)$ of the lots containing p_1 proportion of the defective; and a proportion β of the lots containing p_2 proportion of the defective. Using the above graph one can find how the proportion of the defective are distributed between the upper and lower limits by reading off X for $p = p_1$ and then using two terms on the R.H.S. of (1.13.15) separately. Generally, the larger proportion of the defective p_2 is affected very little by the presence of double limits. Suppose, the proportions are given by

$$\begin{aligned} p_1 &= p_1' + p_1'' \\ \text{and } p_2 &= p_2' + p_2'', \end{aligned}$$

and also assume $p_1' > p_1''$ and $p_2' > p_2''$ so that p_2'' is usually negligible. With this, a sampling plan may be set up by combining the two specification plans while using the parameters (α, p_1') and (α, p_2') at each end. The OC curve may also be obtained from the single specification formula by using the equation (1.13.15), i.e., by transforming the proportion of the defective.

Example 4 *To obtain L and the sample size in a variable sample inspection plan on the basis of sample mean of measurements when population variance is known.*

Solution: Here, items with measurements less than L are not acceptable.

$$P(\bar{X} < L \mid \mu_p) = p_p = \text{Producer's Risk}$$

$$\Rightarrow P\left(\frac{\bar{X}-\mu_p}{\sigma/\sqrt{n}} \leq \frac{L-\mu_p}{\sigma/\sqrt{n}}\right) = p_p,$$

where μ_p is the producer's mean value of measurement and $P(\bar{X} > L \mid \mu_c) = p_c =$ Consumer's Risk

$$\Rightarrow P\left(\frac{\bar{X}-\mu_c}{\sigma/\sqrt{n}} > \frac{L-\mu_c}{\sigma/\sqrt{n}}\right) = p_c,$$

where μ_c is the consumer's mean value of measurement.

Now

$$\frac{\bar{X}-\mu_p}{\sigma/\sqrt{n}} \sim N(0,1) \text{ under } H_0: \ \mu = \mu_p.$$

Similarly,

$$\frac{\bar{X}-\mu_c}{\sigma/\sqrt{n}} \sim N(0,1) \text{ under } H_0: \mu = \mu_c.$$

Denoting both the standard normal variants as τ, we have

$$P(\tau \le e) = p_p$$
$$\text{and } P(\tau > f) = p_c,$$

where

$$e = \frac{L-\mu_p}{\sigma/\sqrt{n}} \text{ and } f = \frac{L-\mu_c}{\sigma/\sqrt{n}}.$$

Since p_p, p_c are known, e and f can be obtained and given μ_p, μ_c and σ, one can get lower specification limit L and the sample size n.

Example 5 *To obtain U, the upper specification limit and the sample size n in a variable sampling inspection plan when population variance σ^2 is known.*

Solution: Here, items with measurements greater than U are not acceptable.

$$P(\bar{X} > U \mid \mu_p) = p_p = \text{Producer's Risk}$$

$$\Rightarrow P\left(\frac{\bar{X}-\mu_p}{\sigma/\sqrt{n}} > \frac{U-\mu_p}{\sigma/\sqrt{n}}\right) = p_p$$

and

$$P(\bar{X} \le U \mid \mu_c) = p_c = \text{Consumer's Risk}$$

$$\Rightarrow P\left(\frac{\bar{X}-\mu_c}{\sigma/\sqrt{n}} \le \frac{U-\mu_c}{\sigma/\sqrt{n}}\right) = p_c.$$

Putting

$$e' = \frac{U-\mu_p}{\sigma/\sqrt{n}} \text{ and } f' = \frac{U-\mu_c}{\sigma/\sqrt{n}}$$

$$P(\tau > e') = p_p \text{ and } P(\tau \le f') = p_c,$$

where τ is a standard normal variate, and p_p and p_c are known which gives e' and f'. Given μ_p, μ_c and σ, one can find U and n.

Remarks

If in Examples 5 or 6, σ remains unknown one has to use t-distribution instead of standard normal variate by getting an unbiased estimate of σ^2, viz.,

$$\frac{1}{n-1}\sum_{i=1}^{n}(X_i - \bar{X})^2.$$

However, in this case under $H_0: p = p_p$; if the distribution of t is central, then under $H_0: p = p_c$, the distribution will be non-central.

Example 6 *Suppose we draw a random sample of size n from the lot and calculate the sample standard deviation s' on the basis of $(n-1)$ degrees of freedom. If $s' > s$ we reject the lot; otherwise we accept the same. Obtain s and n, given p_p (producer's risk), p_c (consumer's risk), σ_p (producer value of standard deviation) and σ_c (consumer value of standard deviation).*

Solution: In this case,

$$P(s' > s \mid \sigma_p) = p_p \Rightarrow P\left(\frac{(n-1)s'^2}{\sigma_p^2} > \frac{(n-1)s^2}{\sigma_p^2}\right) = p_p,$$

where

$$\frac{(n-1)s'^2}{\sigma_p^2} = \chi_{n-1}^2 \sim \text{central } \chi^2 \text{ with } (n-1) \text{ degrees of freedom.}$$

Let

$$\frac{(n-1)s^2}{\sigma_p^2} = \chi_p^2.$$

Hence $P(\chi_{n-1}^2 > \chi_p^2) = p_p$ and similarly, $P(\chi_{n-1}^2 > \chi_c^2) = p_c$, where $P(s' \leq s \ \sigma_c) = p_c$. Thus we have

$$\frac{(n-1)s^2}{\sigma_p^2} = \chi_p^2 \text{ and } \frac{(n-1)s^2}{\sigma_c^2} = \chi_c^2$$

and (n, s) are obtainable from the two relations.

Remarks

If the sample size $n > 30$, then we apply normal approximation to χ^2 as

$$\tau = \sqrt{2\chi_{n-1}^2} - \sqrt{2(n-1)-1} = \sqrt{2\chi_{n-1}^2} - \sqrt{2n-3} \tag{i}$$

so that

$$P(\chi_{n-1}^2 > \chi_p^2) = p_p \tag{ii}$$

$$\Rightarrow \left(\sqrt{2\chi_{n-1}^2} - \sqrt{2n-3} > \sqrt{2\chi_p^2} - \sqrt{2n-3}\right) = p_p, \text{ or, } P(\tau > j) = p_p; \tag{iii}$$

where

$$j = \sqrt{2\chi_p^2} - \sqrt{2n-3}. \tag{iv}$$

Also

$$P(\chi_{n-1}^2 > \chi_c^2) = 1 - p_c \tag{v}$$

$$\Rightarrow P\left(\sqrt{2\chi_{n-1}^2} - \sqrt{2n-3} \geq \sqrt{2\chi_c^2} - \sqrt{2n-3}\right) = 1 - p_c$$

$$\Rightarrow P(\tau \geq k) = 1 - p_c; \text{ where } \tau = \sqrt{2\chi_n^2} - \sqrt{2n-3}. \tag{vi}$$

From (*i*) through (*vi*) we get the value of n and s.

Example 7 *We draw a random sample of size n from the lot and calculate the sample coefficient of variation ν' as $\nu' = s'/\bar{X}'$; where s' = sample standard deviation based on $(n-1)$ degrees of freedom and $\bar{X}'$ is the sample mean.*

Solution: If $\nu' > \nu$, we reject the lot and if $\nu' \leq \nu$, we accept the lot. Obtain ν and n from p_p, p_c, ν_p (Producer's value of coefficient of variation) and ν_c (Consumer's value of coefficient of variation).

We have,

$$P(\nu' > \nu \mid \nu_p) = p_p \Rightarrow P\left[\left(\frac{n\nu'^2}{1+\nu'^2}\right)\left(\frac{1+\nu_p^2}{\nu_p^2}\right) > \left(\frac{n\nu^2}{1+\nu^2}\right)\left(\frac{1+\nu_p^2}{\nu_p^2}\right)\right] = p_p.$$

Now

$$\left(\frac{n\nu'^2}{1+\nu'^2}\right)\left(\frac{1+\nu_p^2}{\nu_p^2}\right) = \chi^2_{n-1}$$

is distributed as central χ^2 with $(n-1)$ distribution function.

Let

$$\left(\frac{n\nu^2}{1+\nu^2}\right)\left(\frac{1+\nu_p^2}{\nu_p^2}\right) = \chi^2_p, \text{ so that } P(\chi^2_{n-1} > \chi^2_p) = p_p. \tag{i}$$

Now consumer's risk is given by

$$P\left[\left(\frac{n\nu'^2}{1+\nu'^2}\right)\left(\frac{1+\nu_c^2}{\nu_c^2}\right) < \left(\frac{n\nu^2}{1+\nu^2}\right)\left(\frac{1+\nu_c^2}{\nu_c^2}\right)\right] = p_c$$

and

$$\left(\frac{n\nu'^2}{1+\nu'^2}\right)\left(\frac{1+\nu_c^2}{\nu_c^2}\right) = \chi^2_{n-1}$$

is a central χ^2 with $(n-1)$ distribution function and

$$\left(\frac{n\nu^2}{1+\nu^2}\right)\left(\frac{1+\nu_c^2}{\nu_c^2}\right) = \chi^2_c \tag{ii}$$

so that

$$\begin{aligned} P(\chi^2_{n-1} < \chi^2_c) &= p_c \\ \text{and } P(\chi^2_{n-1} \geq \chi^2_c) &= 1 - p_c. \end{aligned}$$

Now from (*i*) and (*ii*) we have, by eliminating ν,

$$\frac{\chi^2_p}{\chi^2_c} = \frac{(1+\nu_p^2)/\nu_p^2}{(1+\nu_c^2)/\nu_c^2} = \frac{\nu_c^2(1+\nu_p^2)}{\nu_p^2(1+\nu_c^2)}. \tag{iii}$$

Now R.H.S. of (*iii*) is completely known, as ν_p and ν_c are given. Therefore, by inspecting the column needed by probabilities p_p and $(1-p_c)$ of the χ^2-table we can find the distribution function $(n-1)$ by trial and error so that the ratio (*iii*) is satisfied. Hence n can be found. Then ν can be obtained from either (*i*) or (*ii*).

Remarks

For n > 30 we can apply large sample approximation to χ^2 and hence

$$\sqrt{2\chi^2_{n-1}} - \sqrt{2n} = \tau$$

is distributed as N(0, 1). Thus we modify the process exactly as in the case of sampling inspection plan based on S.D. to get ν and n.

1.14 Single Specification Limit when σ is unknown

As discussed earlier, our problem becomes much more complicated if σ is unknown. Either sample standard deviation or sample range may be used to estimate σ. Duncan (1959) and Owen (1967) may serve as good references to start with the work in this direction. However, we introduce the problems and their solutions in simple form as follows:

Suppose, we have a single lower specification L and the criterion of rejection is $\bar{X} < L$. In other words, if

$$\frac{\bar{X} - L}{s} > k,$$

then we accept the lot, where s is the sample standard deviation and k is suitably chosen number to be decided. In this problem, our object is to find n (the sample size) and k. Let the producer and the consumer values be p_1 and p_2 respectively. The OC curve criterion is to accept the lots having p_1 proportion of the defective with probability $(1 - \alpha)$, i.e., reject the same with probability α, and the lots having p_2 proportion of the defective being accepted with probability β. Under this condition, the probability distribution $(\bar{X} - L)/s$ corresponds to a non-central t, i.e., under the OC curve requirements (α, p_1) and (β, p_2), one has to use table of non-central t distribution (ex. Resinoff *et al.*, 1957). However, a simple approximate procedure is to consider $(\bar{X} - ks)$ and variance

$$\sigma^2\left(\frac{1}{n} + \frac{k^2}{2n}\right),$$

the large sample standard error of s being $\frac{\sigma}{\sqrt{2n}}$.

Then we have

$$P(\bar{X} - ks \leq L) = (1 - \alpha) \text{ under } \left(\frac{\mu - L}{\sigma}\right) = z_{p_1}$$

and

$$P(\bar{X} - ks \leq L) = \beta \text{ under } \left(\frac{\mu - L}{\sigma}\right) = z_{p_2},$$

i.e.,

$$P\left(\frac{\bar{X} - ks - (\mu - k\sigma)}{\sqrt{\sigma^2(1/n + k^2/2n)}} > \frac{L - (\mu - k\sigma)}{\sqrt{\sigma^2(1/n + k^2/2n)}}\right) = (1 - \alpha), \tag{1.14.1}$$

when $\left(\frac{\mu - L}{\sigma}\right) = z_{p_1}$ and

$$P\left(\frac{\bar{X}-ks-(\mu-k\sigma)}{\sqrt{\sigma^2(1/n+k^2/2n)}} \leq \frac{L-(\mu-k\sigma)}{\sqrt{\sigma^2(1/n+k^2/2n)}}\right)=\beta \tag{1.14.2}$$

when $\left(\frac{\mu-L}{\sigma}\right)=z_{p_2}$.

Putting $z_{p_1}\sigma+L=\mu$ in (1.14.1) and $z_{p_2}\sigma+L=\mu$ in (1.14.2) we get,

$$P\left(\frac{\bar{X}-ks-(\mu-k\sigma)}{\sqrt{\sigma^2(1/n+k^2/2n)}} > \frac{-(z_{p_1}-k)}{\sqrt{(1/n+k^2/2n)}}\right)=(1-\alpha) \tag{1.14.3}$$

and

$$P\left(\frac{\bar{X}-ks-(\mu-k\sigma)}{\sqrt{\sigma^2(1/n+k^2/2n)}} \leq \frac{-(z_{p_2}-k)}{\sqrt{1/n+k^2/2n)}}\right)=\beta \tag{1.14.4}$$

Let

$$-z_\alpha = \frac{-(z_{p_1}-k)}{\sqrt{(1/n)+k^2/2n)}} \tag{1.14.5}$$

$$\text{and } z_\beta = \frac{-(z_{p_2}-k)}{\sqrt{(1/n)+k^2/2n)}} \tag{1.14.6}$$

$$\Rightarrow (z_{p_1}-k) = z_\alpha\sqrt{(1/n+k^2/2n)} \tag{1.14.7}$$

$$\text{and } (z_{p_2}-k) = z_\beta\sqrt{(1/n+k^2/2n)}. \tag{1.14.8}$$

Solving (1.14.7) and (1.14.8) for k and n we have

$$k = \frac{z_\beta z_{p_1}+z_\alpha z_{p_2}}{z_\beta+z_\alpha} \tag{1.14.9}$$

$$\text{and } n = \frac{(z_\alpha+z_\beta)^2}{(z_{p_1}\ z_{p_2})^2}(1+k^2/2). \tag{1.14.10}$$

1.15 Double Specification Limit when σ is unknown

In this, of course it is difficult to arrive at an easy solution as was done in the case when σ was known. The main difficulty one may apprehend that the usual approach based on

$$\frac{\bar{X}-L}{s} \geq k \text{ and } \frac{U-\bar{X}}{s} \leq k$$

may not have a unique OC function. In fact, we have a band of OC curves but detailed calculation has revealed that the band is quite narrow. The other procedure is to estimate the minimum percentage defective about

$$2\Phi\left[\frac{(L-U)}{2s}\right] \text{ as in 1.13.2}$$

so that the procedures must reject the lot if s is large. Interested readers are referred to Duncan (1959).

1.16 Dodge's Continuous Sampling Inspection Plan

The plan is applicable to characteristics subject to non-destructive inspection on a 'Go or no Go' basis. It is intended primarily for use in process inspection parts or a final inspection of finished articles subject to the condition that the percentage of defective units in the accepted products will be negligibly low and will not be above a minimum stipulated level.

In operation, the plan provides a corrective sampling procedure of inspection serving as a partial screening for defective cases while inspecting a continuous flow of individual units and is based on the random order of defective units of products, where quality is statistically controlled.

1.16.1 Inspection of the flow of individual units

We consider first the inspection of a single characteristic (i.e., detecting only one kind of defect) in a continuous flow of individual units in the order of production. Assuming that the inspection is to be done for only one characteristic so that attention will be centered in detecting one kind of defect only.

Dodge's AOQL plan is a corrective sampling plan. Dodge has developed this acceptance/rectification plan for continuous production; because when production is continuous, the formation of inspection lots for lot acceptance is somewhat artificial. Moreover, where conveyor lines are used it may be impractical or unduly costly to form inspection lots (i.e., lot sampling procedure).

1.16.2 Procedure of Dodge's inspection plan

- At the outset 100% of the consecutive units are inspected until i units (a stipulated number) in succession are found to be free from defects.
- When i units in succession are found to be free from defects, discontinue 100% inspection and inspect only a fraction f of the units (called *sampling fraction*), selecting the individual units in such a manner that the sample is representative of the population.
- If a single unit is found defective, then we revert to 100% inspection of succeeding units again and the same procedure is continued until a run of i number of non-defective units in succession is recorded.
- This process is continued and all the defective found in the process are corrected or replaced.

1.17 Mathematical Background of Dodge's Inspection Plan

Let the probability of one unit be defective $= p$ and non-defective $= q = (1-p)$. Then the probability of having a terminal defect sequence of n units, i.e., a sequence of $(n-1)$ successive non-defective units followed by a defective nth unit is $q^{n-1}p$. Total probability of all such sequence is

$$\sum_{n=1}^{\infty} q^{n-1}.p = 1.$$

Therefore, the probability of having a failure sequence, i.e., a terminal defective sequence, of i or less units is

$$\sum_{n=1}^{i} q^{n-1}p = 1 - q^i = P, \ (say). \qquad (1.17.1)$$

Probability of having i number of non-defective units in succession = probability of having a terminal defective sequence of more than i units

$$= q^i = \sum_{n=i+1}^{\infty} q^{n-1}p = 1 - (1 - q^i) = Q, \ (say). \qquad (1.17.2)$$

Hence, probability of having j number of failure sequence before getting i units clear of defects $= P^jQ$. Also total probability of all such failure sequence before getting i units free from defects

$$= \sum_{j=0}^{\infty} P^jQ = 1.$$

With these probability statements, let U = expected number of units inspected during complete enumeration $= GH + i$, where G = expected number of failure sequences encountered before getting i units free from defects and H = expected number of units inspected in a failure sequence.

Then,

$$\begin{aligned} G &= \sum_{j=0}^{\infty} jP^jQ = PQ + 2P^2Q + 3P^2Q + \ldots \\ &= PQ(1 + 2P + 3P^2 + \ldots) = PQ(1 - P)^{-2} \\ &= PQ/Q^2 = P/Q = (1 - q^i)/q^i \end{aligned} \qquad (1.17.3)$$

$$\begin{aligned} \text{and } H &= \frac{\sum_{n=1}^{i} nq^{n-1}\, P}{\sum_{n=1}^{i} q^{n-1}p} = \frac{p}{1-q^i}\sum_{n=1}^{i} nq^{n-1}, \text{ by (1.17.1)} \\ &= \frac{p}{1-q^i}\sum_{n=1}^{i}\frac{d}{dq}q^n = \frac{p}{1-q^j}\frac{d}{dq}\sum_{n=1}^{i} q^n. \end{aligned}$$

Now

$$\begin{aligned} \sum_{n=1}^{i} q^n &= q + q^2 + \cdots + q^i = \frac{q(1-q^i)}{1-q}. \\ \therefore \frac{d}{dq}\sum_{n=1}^{i} q^n &= \frac{(1-q)(1-(i+1)q^i) - q(1-q^i)(-1)}{(1-q)^2} \\ &= \frac{1-(1+1)q^i - q + (i+1)q^{i+1} + q - q^{i+1}}{(1-q)^2} \\ &= \frac{1-(i+1)q^i + iq^{i+1}}{p^2} = \frac{1-q^i(i+1-iq)}{p^2} \\ &= \frac{1-q^i(1+ip)}{p^2}. \end{aligned}$$

Also

$$H = \frac{p}{(1-q^i)p^2}[1 - q^i(1+ip)] = \frac{1}{p(1-q^i)}[1 - q^i(1+ip)] = \frac{1-q^i-ipq^i}{p(1-q^i)} \tag{1.17.4}$$

$$\Rightarrow \quad GH = \frac{(1-q^i)(1-q^i-ipq^i)}{q^i p(1-q^i)} = \frac{1-q^i-ipq^i}{pq^i}. \tag{1.17.5}$$

$$\therefore\ U = GH + i = \frac{1-q^i}{pq^i}, \tag{1.17.6}$$

which is the expected number of units during complete enumeration just preceding to the stage of sampling.

Now, let us estimate the average number of units inspected during sampling till we go back to complete enumeration while getting the trace of a single defective unit. Let the expected number of units passed under the sampling inspection procedure be ν (only f proportion of them having been inspected), then $f\nu = h$ = expected number of units inspected during sampling inspection. This is also given by

$$\sum_{n=1}^{\infty} nq^{n-1}p = 1/p.$$

$$\therefore\ \nu = 1/pf. \tag{1.17.7}$$

Average fraction of the total units inspected in the long run

$$F = \frac{U+fV}{U+V},$$

where

$$U = GH + i = \frac{1-q^i}{pq^i} \text{ and } f = 1/pV.$$

Hence Average Outgoing Quality (AOQ) $= p(1-F) + 0 \times F$, where p is the incoming fraction defective and $(1-F)$ represents the proportion of products not having been inspected. Therefore, we have,

$$\text{AOQ} = p_A = p(1-F) = p\left(1 - \frac{U+fV}{U+V}\right) = p\frac{\nu(1-f)}{U+V}.$$

Substituting from (1.17.6) to (1.17.7), we have

$$\begin{aligned}\text{AOQ} = p_A &= p\frac{(1-f)/pf}{[(1-q^i)/pq^i] + 1/pf} = p\frac{(1-f)/f}{[(1-q^i)/q^i] + 1/f} \\ &= pq^i\frac{(1-f)}{f(1-q^i)+q^i} = pq^i\frac{(1-f)}{f+(1-f)q^i}\end{aligned} \tag{1.17.8}$$

$$\Rightarrow p_A f + (1-f)p_A q^i = pq^i(1-f). \tag{1.17.9}$$

AOQL is obtained from $dp_A/dp = 0$, after some simplification in (1.17.8)

$$\Rightarrow f(1-p) - ipf + (1-f)(1-p)^{i+1} = 0$$

$$\Rightarrow 1 + \frac{(1-f)}{f}(1-p)^i = \frac{ip}{(1-p)}. \tag{1.17.10}$$

Let the solution of (1.17.10) be $p = p_1$. We then go on taking different levels of p, the AOQL and use Dodge's chart to find f and i for different levels of AOQL. This is graphically represented in the following example.

Dodge gives a chart for the ready application of this inspection plan for any desired AOQL. This is obtained by plotting the points (i, f) under different levels of AOQL; each AOQL giving a particular value of i for a given f (or a particular value of f for a given i). This is shown in Fig 7 below.

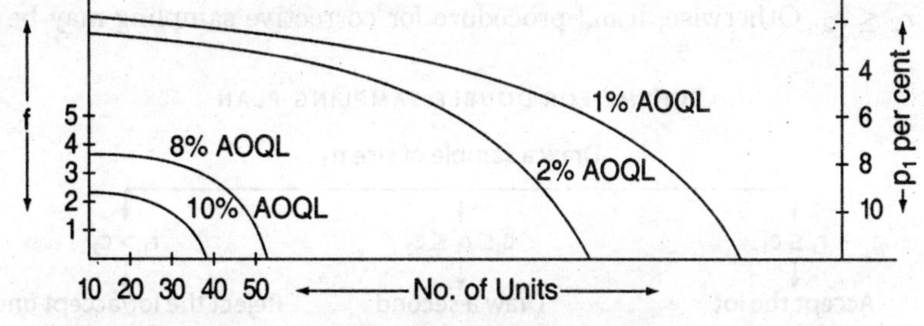

Fig 7 Curves for determining f (or i) at given level of AOQL

As an example, suppose the desired AOQL is 2% and it is proposed to establish a plan that calls for inspection of one piece out of every 20 pieces from the conveyor's belt. Then f, the fraction inspected is 1/20, i.e., 5%. Using the above figure, we can find the value of i corresponding to $f = 5\%$ and AOQL $= 2\%$. This gives $i = 76$. The acceptance is then as follows:

- Inspect all the units consecutively as produced until 76 units in succession are free from defects. As soon as 76 successive units are free from defect, inspect a sample consisting of one unit out of 20. Accept all the products as long as the sample is free from defects.
- Whenever one of these sample units is found to be defective resume 100% inspection again until 76 units in succession have been found free from defects. After the realisation of the same, go back to sampling inspection.

The 2% AOQL for this plan assumes that defective pieces occur at random. This limitation may be made clear by assuming that after the plan had started and the inspection was on sampling basis, some fault might have crept in the production process causing perhaps eighth, ninth or tenth units out of every twenty units defective; without affecting other units. However, this kind of limitations in respect of AOQL is not of much practical importance because such cyclical causes in continuous production are rare. Even if there is any apprehension that such possibilities may exist, often the fact becomes evident from the physical process of production. Nevertheless, it should be recognised that the present acceptance does not protect against circumstances under which the defective may be dishonestly manipulated to fall in the intervals between regularly spaced samples.

1.18 Double Sampling Inspection Plan

Under single sampling plan, decision in regard to a lot is taken on the basis of a criterion based on a single sample. As pointed out earlier, chance factors play a lot to decide the outcome as the decision is based on the outcome of a single experiment. The idea of a double sampling plan is to eliminate the effect of chance factor partially, of course at the expense of increasing the cost of sampling inspection by deciding the fate of a lot (acceptance or rejection) on the basis of two trials. The procedure is given as follows:

First we draw a sample of size n_1 from a lot of size N and let r_1 be the number of the defective out of it. If $r_1 \leq c_1$ (a given number), then we accept the lot. If $r_1 > c_2$, then we either reject the lot or accept the same on the basis of corrective sampling plan (with or without replacement), where c_2 is another chosen number greater than c_1.

If, however, $r_1 > c_1$ but $\leq c_2$, then draw a second sample of size n_2 and let r_2 be the number of the defective in it. If $r_1 + r_2 > c_2$, then reject the lot and accept it under acceptance sampling plan if $r_1 + r_2 \leq c_2$. Otherwise, usual procedure for corrective sampling may be followed.

SCHEME FOR DOUBLE SAMPLING PLAN

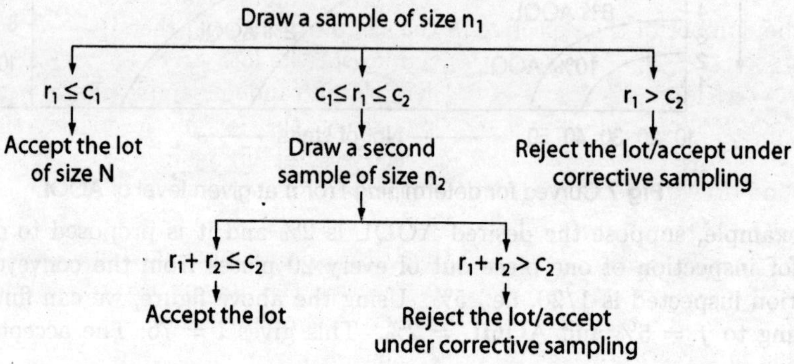

Consumer's Risk $= P$ [the lot is accepted $|p = p_c =$ consumer's value of p]

$$= \sum_{r_1=0}^{c_1} \binom{n_1}{r_1} p_c^{r_1} (1-p_c)^{n_1-r_1} + \sum_{r_1=c_1+1}^{c_2} \binom{n_1}{r_1} p_c^{r_1} (1-p_c)^{n_1-r_1} \times \sum_{r_2=0}^{c_2-r_1} \binom{n_2}{r_2} p_c^{r_2} (1-p_c)^{n_2-r_2}. \tag{1.18.1}$$

Producer's Risk $= 1 - P$ [the lot is accepted $|p = p_p =$ producer's value of p]

Now

$$\text{OC function} = P\,[\text{the lot is accepted}] \tag{1.18.2}$$

$$= \sum_{r_1=0}^{c_1} \binom{n_1}{r_1} p^{r_1} (1-p)^{n_1-r_1} + \sum_{r_1=c_1+1}^{c_2} \binom{n_1}{r_1} p^{r_1} (1-p)^{n_1-r_1} \times \sum_{r_2=0}^{c_2-r_1} \binom{n_2}{c_2} p^{r_2} (1-p)^{n_2-r_2}. \tag{1.18.3}$$

ASN (under acceptance sampling plan) $= n_1 + n_2 P[c_1 \leq r_1 \leq c_2]$

$$= n_1 + n_2 \sum_{r_1=c_1+1}^{c} \binom{n_1}{r_1} p^{r_1} (1-p)^{n_1-r_1}. \tag{1.18.4}$$

ASN (under corrective sampling plan)

$$= n_1 + (N - n_1)P[r_1 > c_2] + n_2P[c_1 \leq r_1 \leq c_2] + (N - n_1 - n_2)P[c_1 \leq r_1 \cap r_2 > c_2 - r_1] \quad (1.18.5)$$

1.19 Multiple Sampling Plan

Let $c_1 \leq c_1', c_2 \leq c_2', \ldots, c_{s-1} \leq c_{s-1}', c_s \leq c_s'$ be s pairs of numbers. Draw a random sample of size n_1.

If r_1 is the number of the defective, then if

$r_1 \leq c_1$;	we accept the lot.
$r_1 > c_1$ but $\leq c_1'$;	then we draw another sample of size n_2.
$r_1 > c_1'$;	we reject the lot or accept it with the principle of corrective sampling plan.

If r_2 is the number of the defective in the sample of size n_2, then if

$r_1 + r_2 \leq c_2$;	then accept the lot.
$r_1 + r_2 > c_2 \leq c_2'$;	then draw a random sample of size n_3.
$r_1 + r_2 > c_2'$;	then reject the lot or accept on the basis of corrective sampling plan.

Suppose, on drawing a random sample of size n_3, if

$r_1 + r_2 + r_3 \leq c_3$;	then we accept the lot.
$r_1 + r_2 + r_3 > c_3 \leq c_3'$;	then draw a random sample of size n_4.
$r_1 + r_2 + r_3 > c_3'$;	then reject the lot or accept on the basis of corrective sampling plan.

Proceeding in this way, at the $(s-1)$th stage, if we have

$r_1 + r_2 + \cdots + r_{s-1} \leq c_{s-1}$;	then we accept the lot.
$r_1 + r_2 + \cdots + r_{s-1} > c_{s-1} \leq c_{s-1}'$;	draw another sample of size n_s.
$r_1 + r_2 + \cdots + r_{s-1} > c_{s-1}'$,	reject the lot or accept it on the basis of corrective sampling plan.

Finally, if

$r_1 + r_2 + \cdots + r_s \leq c_s$;	then we accept the lot.
$r_1 + r_2 + \cdots + r_s > c_s$;	reject the lot or accept under the criterion of corrective sampling plan.

NOTE

We do not allow the process to continue indefinitely. At the sth stage if $r_1 + r_2 + \cdots + r_s > c_s$ and may be $\leq c_s'$ for some number $c_s' > c_s$. But since we deliberately plan to take a decision at sth stage no c_s' has been kept out for this purpose.

1.20 Sequential Probability Ratio Test

Multiple sampling plan as discussed in the earlier section may further be extended to sequential sampling inspection plan by employing the principle of Sequential Probability Ratio Test (SPRT). This operates for acceptance sampling as follows:

- To inspect a sample from a lot at random.

- On the result of inspection leading to acceptance, rejection or continue taking further observations, based on certain specific rules, dictated by the procedure of SPRT are as follows:

If at the mth stage of sampling ($m = 1, 2, 3, \ldots$) we are to a choose two numbers A_m and B_m, known as *acceptance* and *rejection numbers* respectively, in the sense that if up to mth stage, the total number of defective pieces be less than or equal to A_m, then the lot is accepted; if greater than B_m, the lot is rejected and if it is greater than A_m but less than B_m one more sample of observations is to be taken and another decision is to be made with the help of two other similar numbers A_{m+1} and B_{m+1} at the $(m+1)$th stage. The process is continued till a final decision is taken in regard to the acceptance or rejection of the lot.

Remarks

The procedure of sequential sampling inspection plan can be taken as a stochastic generalisation of the deterministic multiple sampling inspection plan. In the former, the stopping rule is given by a probabilistic consideration and it can be proved that SPRT procedure of drawing samples and then taking decisions terminates with probability one. Whereas in the latter case, the stopping stage is prefixed and deliberate.

1.20.1 The basis of sequential probability ratio test as applied to variables

Let X be a random variable whose probability density function is $f(x, \theta)$, θ being an unknown parameter. We require to get information on θ, *say,* in the form of testing.

$$H_0 : \theta = \theta_0 \text{ (producer's value of the proportion of the defective)}$$

$$\text{against} \quad H_1 : \theta = \theta_1 \text{ (consumer's value of the proportion of the defective).}$$

Let α and β be the errors of type I (E_I) and type II (E_{II}) respectively. Wald has given the basis of sequential sampling plan which operates as follows:

Let $X_1, X_2, \ldots, X_m$ be the random sample of observations, where the number is not fixed but a random variable (r.v.). The probability of getting such a sample is

$$P_m = \prod_{i=1}^{m} f(x_i, \theta);$$

since the observations are independent in a random sample.

Let p_{1m} and p_{0m} be the values of n when $\theta = \theta_1$ and $\theta = \theta_0$ respectively.

If, for two specially chosen numbers A and B,

$$\frac{p_{1m}}{p_{0m}} \geq A, \text{ then } H_0 : \theta = \theta_0 \text{ will be rejected against } H_1 : \theta = \theta_1$$

and if

$$\frac{p_{1m}}{p_{0m}} \leq B, \text{ then } H_0 : \theta = \theta_0 \text{ will be accepted against } H_1 : \theta = \theta_1.$$

On the other hand, if

$$B < \frac{p_{1m}}{p_{0m}} < A,$$

the result ceases to be conclusive and we draw further observations.

Again

$$\frac{p_{1m}}{p_{0m}} \geq A \Rightarrow P(E \in w \mid H_1) \geq AP(E \in W \mid H_0),$$

where $E \equiv (x_1, x_2, \ldots, x_n)$, m dimensional sample space,
$W \equiv$ sample space and
$w \equiv$ critical region,

$$\text{i.e., } (1-\beta) \geq A\alpha \Rightarrow A \leq \frac{1-\beta}{\alpha}. \tag{1.20.1}$$

Precisely with the same reasoning

$$\frac{p_{1m}}{p_{0m}} \leq B \Rightarrow P(E \in W - w \mid H_1) \leq BP(E \in W - w \mid H_0)$$

$$\beta \leq B(1-\alpha) \Rightarrow B \geq \frac{\beta}{1-\alpha}. \tag{1.20.2}$$

Despite the inequalities in (1.20.1) and (1.20.2), we take on approximation,

$$A \cong \frac{1-\beta}{\alpha} \text{ and } B \cong \frac{\beta}{1-\alpha}, \tag{1.20.3}$$

assuming the deficit of A below $\frac{1-\beta}{\alpha}$ and the excess of B over $\frac{\beta}{1-\alpha}$, balancing each other in the long run. Of course, with this approximation, α and β cease to be the size of the type I error and type II error respectively. Moreover, if α' and β' represent the size of the type I error and type II error respectively, then

$$\frac{1-\beta'}{\alpha'} \geq \frac{1-\beta}{\alpha} \text{ and } \frac{\beta'}{1-\alpha'} \leq \frac{\beta}{1-\alpha}$$
$$\Rightarrow \quad \alpha' \leq \alpha\frac{1-\beta'}{1-\beta} \text{ and } \beta' \leq \beta\frac{1-\alpha'}{1-\alpha}.$$

In other words, $\alpha' < \alpha \Rightarrow \beta' > \beta$.

1.21 OC Function of the SPRT

Let us test $H_0 : \theta = \theta_0$ when θ is an attribute given in terms of a proportion (*say*, the proportion of defective bulbs in a manufactured lot). Then

$$L(\theta_0) = P[E \in (W-w) \mid H_0] = 1 - P(E \in w \mid H_0) = 1-\alpha,$$

where

$$P(E \in w \mid H_0) = \alpha, \text{ size of the type I error}$$

and

$$L(\theta_1) = P[E \in (W-w) \mid H_1] = \beta,$$

size of the type II error.

Again $L(0) = 1$ and $L(1) = 0$, obviously.

Therefore, besides the OC function defined at $\theta_0, \theta_1, 0$ and 1 respectively as

$$L(\theta_0) = 1 - \alpha,\ L(\theta_1) = \beta,\ L(0) = 1 \text{ and } L(1) = 0$$

and there is a fifth point known as *special point* which corresponds to some function of θ, *say,* $h(\theta)$ subject to the condition

$$E(R^{h(\theta)}) \equiv 1, \text{ where } R = \frac{f(x,\theta_1)}{f(x,\theta_0)}.$$

$$E(R^{h(\theta)}) = 1 \quad \Rightarrow \quad \int \left(\frac{f(x,\theta_1)}{f(x,\theta_0)}\right)^{h(\theta)} f(x,\theta)\,dx = 1$$

$$\Rightarrow \quad \left(\frac{f(x,\theta_1)}{f(x,\theta_0)}\right)^{h(\theta)} f(x,\theta) \text{ is a density function } = g(x,\theta), \text{ (say)}.$$

$$\Rightarrow \quad \left(\frac{f(x,\theta_1)}{f(x,\theta_0)}\right)^{h(\theta)} = \frac{g(x,\theta)}{f(x,\theta)}.$$

SPRT remains inconclusive, if

$$B < \frac{f(x,\theta_1)}{f(x,\theta_0)} < A$$

$$\Rightarrow \quad B^{h(\theta)} < \left(\frac{f(x,\theta_1)}{f(x,\theta_0)}\right)^{h(\theta)} < A^{h(\theta)}, \text{ if } h(\theta) \geq 0.$$

Putting $B^{h(\theta)} = B'$ and $A^{h(\theta)} = A'$ we have,

$$B' < \frac{g(x,\theta)}{f(x,\theta)} < A'.$$

But because of transformations B to B' and A to A', sizes of the errors get changed. Let the size of type I error (E_I) and type II error (E_{II}) get changed to α' and β' respectively,

$$\text{i.e.,} \quad B' = \frac{\beta'}{1-\alpha'} \text{ and } A' = \frac{1-\beta'}{\alpha'}$$

$$\Rightarrow \quad B' - \alpha' B' = \beta';\ A'\alpha' = 1 - \beta'$$

$$\Rightarrow \quad B' - \alpha'(B' - A') = 1, \text{ by adding}$$

$$\Rightarrow \quad \frac{B'-1}{B'-A'} = \alpha' = \frac{1-B'}{A'-B'}$$

$$\Rightarrow \quad \alpha' = \frac{1-B^h}{A^h - B^h}.$$

We have,

$$L(\theta) = P(E \in W - w \mid H).$$

$$\therefore\ L(\theta_0) = P(E \in W - w \mid H_0) = 1 - \alpha'.$$

Substituting the value of α', we have

$$L(\theta_0) = 1 - \frac{1 - B^{h(\theta_0)}}{A^{h(\theta_0)} - B^{h(\theta_0)}} = \frac{A^{h(\theta_0)} - B^{h(\theta_0)} - 1 + B^{h(\theta_0)}}{A^{h(\theta_0)} - B^{h(\theta_0)}} = \frac{A^{h(\theta_0)} - 1}{A^{h(\theta_0)} - B^{h(\theta_0)}}$$

$$\Rightarrow \quad L(\theta) = \frac{A^{h(\theta_0)} - 1}{A^{h(\theta)} - B^{h(\theta)}}, \text{ subject to } E\left[\frac{f_1}{f_0}\right]^{h(\theta)} = 1,$$

where f_0 and f_1 are the densities under H_0 and H_1 respectively.

One can now plot the OC function by five points for testing the attribute given by

$$H_0 : \theta = \theta_0 \text{ against } H_1 : \theta = \theta_1.$$

The five points are

$$L(\theta_0) = 1 - \alpha,\ L(\theta_1) = \beta,\ L(0) = 1,\ L(1) = 0$$

and finally,

$$L(\theta) = \frac{A^{h(\theta)} - 1}{A^{h(\theta)} - B^{h(\theta)}}$$

for a special point θ, such that $E\left[\frac{f_1}{f_0}\right]^{h(\theta)} = 1.$

1.21.1 An alternative method for construction of OC function of the attribute

We illustrate the technique with the assumption that the attribute θ is a parameter of the Bernoullian distribution. We require to test the hypothesis $H_0 : \theta = \theta_0$ against the alternative $H_1 : \theta = \theta_1$. We have,

$$\left(\frac{f_1}{f_0}\right) = \frac{\binom{1}{k}\theta_1^k(1-\theta_1)^{1-k}}{\binom{1}{k}\theta_0^k(1-\theta_0)^{1-k}}, \text{ where } k = 0, 1.$$

This gives,

$$\left(\frac{f_1}{f_0}\right)^{h(\theta)} = \left[\left(\frac{\theta_1}{\theta_0}\right)^k \left(\frac{1-\theta_1}{1-\theta_0}\right)^{1-k}\right]^{h(\theta)}.$$

We set

$$E\left[\frac{f_1}{f_0}\right]^{h(\theta)} = 1 \tag{1.21.1}$$

$$\Rightarrow \sum_{k=0}^{1}\left[\left(\frac{\theta_1}{\theta_0}\right)^k \left(\frac{1-\theta_1}{1-\theta_0}\right)^{1-k}\right]^{h(\theta)} \binom{1}{k}\theta^k(1-\theta)^{1-k} = 1$$

$$\Rightarrow \left(\frac{1-\theta_1}{1-\theta_0}\right)^{h(\theta)} - \theta\left(\frac{1-\theta_1}{1-\theta_0}\right)^{h(\theta)} + \theta\left(\frac{\theta_1}{\theta_0}\right)^{h(\theta)} = 1 \tag{1.21.2}$$

$$\Rightarrow \theta = \frac{1 - \left(\frac{1-\theta_1}{1-\theta_0}\right)^{h(\theta)}}{\left(\frac{\theta_1}{\theta_0}\right)^{h(\theta)} - \left(\frac{1-\theta_1}{1-\theta_0}\right)^{h(\theta)}}, \tag{1.21.3}$$

where $h(\theta)$ is unknown. We plot θ against $h(\theta)$ taking a series of hypothetical values of $h(\theta)$ and get the corresponding values of θ using (1.21.3) which satisfies (1.21.1). In practice, to plot the OC function given by

$$L(\theta) = \frac{A^{h(\theta)} - 1}{A^{h(\theta)} - B^{h(\theta)}}$$

satisfying the condition set at (1.21.1) we need not solve $h(\theta)$, but for certain hypothetical values of $h(\theta)$, find the corresponding values of θ using (1.21.3) and by plotting $(\theta, L(\theta))$ we get the OC function.

1.21.2 OC function of the variable

Here our object is to test $H_0: \theta = \theta_0$ against $\theta = \theta_1$, where θ is the mean of normal variable under consideration and its standard deviation is assumed to be known. In particular, this is taken to be unity. Now equation (1.21.4) gives

$$\frac{1}{\sqrt{2\pi}}\int_{-\infty}^{\infty} e^{1/2(x-\theta)^2}\left[\frac{\frac{1}{\sqrt{2\pi}}e^{-1/2(x-\theta_1)^2}}{\frac{1}{\sqrt{2\pi}}e^{-1/2(x-\theta_0)^2}}\right]^{h(\theta)} dx = 1. \tag{1.21.4}$$

We require to obtain $h = h(\theta)$ for which the exponent in the above integrand, viz.,

$$-\frac{(x-\theta)^2}{2} - \frac{h(x-\theta_1)^2}{2} + \frac{h(x-\theta_0)^2}{2}$$

$$= -\frac{1}{2}\left[[x - \{\theta + h(\theta_1-\theta_0)\}]^2 - [\theta + h(\theta_1-\theta_0)]^2\right] - \frac{\theta^2}{2} - h\frac{(\theta_1^2-\theta_0^2)}{2}.$$

$\therefore$ (1.21.4) becomes

$$e^{1/2\{\theta+h(\theta_1-\theta_0)\}^2-\theta^2/2-\frac{h}{2}(\theta_1^2-\theta_0^2)} \times \frac{1}{\sqrt{2\pi}}\int_{-\infty} e^{-1/2[x-\{\theta+h(\theta_1-\theta_0)\}]^2}\, dx = 1.$$

It follows that

$$\frac{1}{2}\{\theta + h(\theta_1-\theta_0)\}^2 = \frac{\theta^2}{2} + \frac{h(\theta_1^2-\theta_0^2)}{2}. \tag{1.21.5}$$

This, on simplification, gives

$$h(\theta) = \frac{\theta_1+\theta_0-2\theta}{\theta_1-\theta_0};\ \theta_1 \neq \theta_0. \tag{1.21.6}$$

1.21.3 Average sampling number under SPRT

SPRT differs from Neyman-Pearsonian (NP) testing in the sense that the sample size is a random variable in case of SPRT; while in NP testing the same is held constant at the expense of keeping one of the error size (usually E_1) fixed. We have for SPRT,

$$\frac{f_1}{f_0} \geq A \Rightarrow \log\frac{f_1}{f_0} \geq \log A \Rightarrow \text{rejection of } H_0$$

Also

$$\frac{f_1}{f_0} \leq B \Rightarrow \log\frac{f_1}{f_0} \leq \log B \Rightarrow \text{rejection of } H_1$$

and

$$B < \frac{f_1}{f_0} < A \Rightarrow \log B < \log\frac{f_1}{f_0} < \log A$$

$\Rightarrow$ no decision regarding acceptance or rejection of either of the hypothesis.

Let us put

$$Z_i = \log \frac{f(x_i, \theta_1)}{f(x_i, \theta_0)},$$

then

$$\Sigma Z_i \geq \log A \Rightarrow \text{rejection of } H_0$$

$$\Sigma Z_i \leq \log B \Rightarrow \text{rejection of } H_1$$

and $\log B < \Sigma Z_i < \log A$ implies that further observations are necessary to draw conclusion.

Z_i's are all independent identically distributed variables since x_i's are so and at the nth observation a decision is taken. Let us take a positive integer N sufficiently large so that $P[n > N] = 0$.

We have

$$(Z_1 + Z_2 + \cdots + Z_N) = (Z_1 + Z_2 + \cdots + Z_n) + (Z_{n+1} + \cdots + Z_N)$$
$$\Rightarrow \quad E_\theta(Z_1 + Z_2 + \cdots + Z_N) = E_\theta(Z_1 + Z_2 + \cdots + Z_n) + E_0(N - n)E_\theta(Z)$$

assuming that Z_i's are independent identically distributed as Z and distribution of n is independent of Z.

$$\Rightarrow \quad NE_\theta(z) = E_\theta(z_1 + z_2 + \cdots + z_n) + NE_\theta(z) - E_\theta(n)E_\theta(z)$$
$$\Rightarrow \quad \text{ASN} \equiv E_\theta(n) = \frac{E_\theta(Z_1 + Z_2 + \cdots + Z_n)}{E_\theta(Z)}, \text{ provided } E_\theta(Z) \neq 0. \qquad (1.21.7)$$

Next to obtain $E_\theta(Z_1 + Z_2 + \cdots + Z_n)$, we note that

$$\log A \leq \log \frac{1-\beta}{\alpha} \text{ and } \log B \geq \log \frac{\beta}{1-\alpha}.$$

Assuming the deficit of $\log A$ over $\log \frac{1-\beta}{\alpha}$ to be of the same order as that of excess of $\log B$ over $\frac{\beta}{1-\alpha}$, we may write

$$E_\theta = (Z_1 + Z_2 + \cdots + Z_n) \cong L(\theta) \log \frac{\beta}{1-\alpha} + [1 - L(\theta)] \log \frac{1-\beta}{\alpha},$$

where $L(\theta)$ represents the probability of accepting the hypothesis. Then the above can be written as

$$E_\theta(n) = \frac{L(\theta) \log \frac{\beta}{1-\alpha} + [1 - L(\theta)] \log \frac{1-\beta}{\alpha}}{E_\theta(Z)} \qquad (1.21.8)$$

provided $E_\theta(Z) \neq 0$.

Special case when $E_\theta(Z) = 0$

Let

$$\sum_{i=1}^{N} Z_i = Z_1 + Z_2 + \cdots + Z_n + \sum_{i=n+1}^{N} Z_i$$

$$E_\theta(S_n^2) = \sum_{i=1}^{N} E_\theta(Z_i^2) - E_\theta(N - n)E_\theta(Z^2),$$

identically; product terms vanish since Z_i's are all independent, where

$$\begin{aligned} S_n &= Z_1 + Z_2 + \cdots + Z_n \\ &= NE_\theta(Z^2) - NE_\theta(Z^2) + E_\theta(n)E_\theta(Z^2) \\ \Rightarrow E_\theta(n) &= \frac{E_\theta(S_n^2)}{E_\theta(Z^2)} \end{aligned} \tag{1.21.9}$$

Again

$$\begin{aligned} E_\theta(S_n^2) &\cong (\log A)^2[1 - L(\theta)] + (\log B)^2 L(\theta) \\ &= \left(\log \frac{1-\beta}{\alpha}\right)^2 [1 - L(\theta)] + \left(\log \frac{\beta}{1-\alpha}\right)^2 L(\theta) \end{aligned} \tag{1.21.10}$$

following the argument leading to the result (1.21.8). Putting (1.21.10) in (1.21.9), we have

$$E_\theta(n) = \frac{\left(\log \frac{1-\beta}{\alpha}\right)^2 [1 - L(\theta)] + \left(\log \frac{\beta}{1-\alpha}\right)^2 L(\theta)}{E_\theta(Z^2)}. \tag{1.21.11}$$

Example 8 *Obtain the* ASN *for testing* $\theta_0 = 0$ *against* $\theta_1 = 1$, *where* θ *is the normal mean with unit variance keeping* $\alpha = 0.01$ *and* $\beta = 0.01$ *(i.e., keeping both producer's risk and consumer's risk at par at* 1% *level).*

Solution: We have

$$\begin{aligned} f(x, \theta \mid \theta) &= \frac{1}{\sqrt{2\pi}} e^{-1/2(x-\theta)^2} \\ Z &= \log \frac{f_1}{f_0} = -\tfrac{1}{2}[(x - \theta_1)^2 - (x - \theta_0)^2] \\ &= \tfrac{1}{2}(\theta_0 - \theta_1)(\theta_0 + \theta_1 - 2x). \\ E(Z) &= \tfrac{1}{2}(\theta_0 - \theta_1)(\theta_0 + \theta_1 - 2\theta) \\ &= \tfrac{1}{2}(\theta_0 - \theta_1)(\theta_0 + \theta_1 - 2\theta). \end{aligned}$$

Hence

$$E_\theta(n) = \frac{2\left[L(\theta)\log \frac{\beta}{1-\alpha} + [1 - L(\theta)]\log \frac{1-\beta}{\alpha}\right]}{(\theta_0 - \theta_1)(\theta_0 + \theta_1 - 2\theta)}$$

►► NOTE

In this case $E_\theta(Z) = 0$ when $\theta = 1/2(\theta_0 + \theta_1)$.

But

$$E\left[\frac{f_1}{f_0}\right]^{h(\theta)} = 1 \Rightarrow \tfrac{1}{2}[\theta + h(\theta_1 - \theta_0)^2] = \tfrac{1}{2}\theta^2 + \frac{h}{2}(\theta_1^2 - \theta_0^2), \textit{ vide } (1.21.5)$$

which gives

$$h(\theta) = \frac{\theta_1 + \theta_0 - 2\theta}{\theta_0 - \theta_1}.$$

$h(\theta)$ will be zero if

$$\theta = \frac{\theta_1 + \theta_0}{2}.$$

Then the OC function is

$$\begin{aligned} \lim_{h\to 0} h(\theta) &= \lim_{h\to 0} \frac{A^h - 1}{A^h - B^h} = \lim_{h\to 0} \frac{A^h \log_e A}{A^h \log_e A - B^h \log_e B} \\ &= \frac{\log A}{\log A - \log B}. \end{aligned} \tag{1.21.12}$$

Hence we consider

$$\begin{aligned} Z^2 &= \tfrac{1}{4} (\theta_0 - \theta_1)^2[(\theta_0 + \theta_1 - 2x)^2] \\ &= \tfrac{1}{4} (\theta_0 - \theta_1)^2[(\theta_0 + \theta_1)^2 - 4(\theta_0 - \theta_1)x + 4x^2]. \\ E(z^2) &= (\theta_0 - \theta_1)^2\theta_1(\theta_0 + \theta_1 + 1). \end{aligned}$$

Then,

$$\begin{aligned} \text{when } h \to 0, \text{ ASN} &= \frac{\dfrac{\log A}{\log A - \log B}(\log B)^2 + \left[1 - \dfrac{\log A}{\log A - \log B}\right](\log A)^2}{(\theta_0 - \theta_1)^2} \\ &= -\frac{\log A \log B}{(\theta_0 - \theta_1)^2} = -\frac{\log\dfrac{1-\beta}{\alpha}\log\dfrac{\beta}{1-\alpha}}{(\theta_0 - \theta_1)^2}. \end{aligned} \tag{1.21.13}$$

With these assigned values, $E_0(Z^2) = (\theta_0 - \theta_1)^2$ on simplification

$$E_\theta(n) = \frac{0.99 \log \dfrac{1}{0.99} + 0.01 \log 0.99}{-1/2} \cong 9.0.$$

1.21.4 Sequential sampling inspection plan

Let the producer's value of fraction defective be π_0 and that of consumer is π_1. We assume $\pi_1 > \pi_0$. Let producer's risk P_p be α and consumer's risk P_c be β. We require to test

$$H_0: \ \pi = \pi_0 \text{ against } H_1: \ \pi = \pi_1$$

with $E_{\text{I}} \equiv$ size of the error of type I $= \alpha$
and $E_{\text{II}} \equiv$ size of the error of type II $= \beta$.

Let a sample of size m be drawn and m^* be the number of the defective found therein. Evidently, the distribution of the number of the defective will be binomial. Hence $p =$ probability of getting m^* defective out of a sample of m items with population fraction defective π,

$$p_m = {}^mC_{m^*}\pi^{m^*}(1 - \pi)^{m-m^*}.$$

Let p_{0m} be the probability under H_0 and p_{1m} under H_1, corresponding to p_m, i.e.,

$$p_{1m} = {}^{m}C_{m^*}\pi_1^{m^*}(1-\pi_1)^{m-m^*}$$

and

$$p_{0m} = {}^{m}C_{m^*}\pi_0^{m^*}(1-\pi_0)^{m-m^*}.$$

Hence

$$\frac{p_{1m}}{p_{0m}} = \left(\frac{\pi_1}{\pi_0}\right)^{m^*}\left(\frac{1-\pi_1}{1-\pi_0}\right)^{m-m^*}$$

$$\begin{aligned}\log\frac{p_{1m}}{p_{0m}} &= m^*\log\frac{\pi_1}{\pi_0} + (m-m^*)\log\frac{1-\pi_1}{1-\pi_0}\\ &= m\log\frac{1-\pi_1}{1-\pi_0} + m^*\left(\log\frac{\pi_1}{\pi_0} - \log\frac{1-\pi_1}{1-\pi_0}\right).\end{aligned}$$

SPRT remains inconclusive if

$$\log\frac{\beta}{1-\alpha} < m\log\frac{1-\pi_1}{1-\pi_0} + m^*\left(\log\frac{\pi_1}{\pi_0} - \log\frac{1-\pi_1}{1-\pi_0}\right) < \log\frac{1-\beta}{\alpha}$$

$$\Rightarrow \frac{\log\dfrac{\beta}{1-\alpha} - m\log\dfrac{1-\pi_1}{1-\pi_0}}{\log\dfrac{\pi_1}{\pi_0} - \log\dfrac{1-\pi_1}{1-\pi_0}} < m^* < \frac{\log\dfrac{1-\beta}{\alpha} - m\log\dfrac{1-\pi_1}{1-\pi_0}}{\log\dfrac{\pi_1}{\pi_0} - \log\dfrac{1-\pi_1}{1-\pi_0}},$$

i.e., $A_m < m^* < R_m$, where

$$A_m = \frac{\log\dfrac{\beta}{1-\alpha} - m\log\dfrac{1-\pi_1}{1-\pi_0}}{\log\dfrac{\pi_1}{\pi_0} - \log\dfrac{1-\pi_1}{1-\pi_0}} \text{ and } R_m = \frac{\log\dfrac{1-\beta}{\alpha} - m\log\dfrac{1-\pi_1}{1-\pi_0}}{\log\dfrac{\pi_1}{\pi_0} - \log\dfrac{1-\pi_1}{1-\pi_0}}.$$

The inequality remains unchanged when $\pi_1 > \pi_0$. Therefore, we continue sampling up to $(m+1)$th stage. Thus we have the test procedure as under:

If $A_m < m^* < R_m$ we continue inspection.

$m^* \leq A_m$ accept H_0 and hence the lot.

$m^* \geq A_m$ reject H_0 and hence the lot.

We can further write

$$\begin{aligned}A_m &= h_0 - ms\\ \text{and } R_m &= h_1 - ms,\end{aligned}$$

where

$$h_0 = \frac{\log\dfrac{\beta}{1-\alpha}}{\log\dfrac{\pi_1}{\pi_0} - \log\dfrac{1-\pi_1}{1-\pi_0}};\ h_1 = \frac{\log\dfrac{1-\beta}{\alpha}}{\log\dfrac{\pi_1}{\pi_0} - \log\dfrac{1-\pi_1}{1-\pi_0}} \text{ and } s = \frac{\log\dfrac{1-\pi_1}{1-\pi_0}}{\log\dfrac{\pi_1}{\pi_0} - \log\dfrac{1-\pi_1}{1-\pi_0}}.$$

This is diagrammatically represented in Fig 8 below.

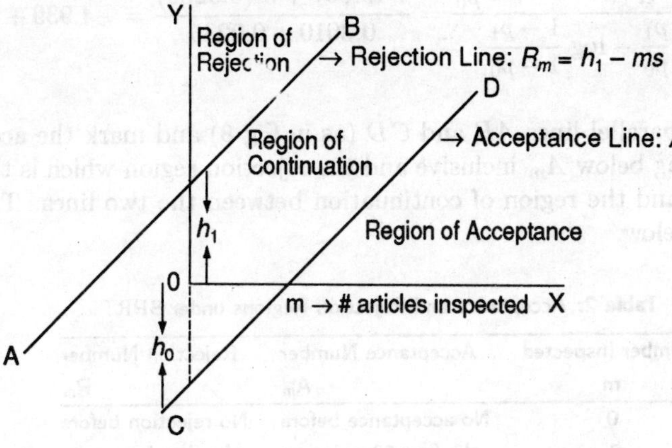

Fig 8 Diagrammatic Representation of Sequential Inspection Based on Fraction Defective

Here for each m we calculate m^* and plot (m, m^*). If the point P lies on or above the line AB we reject the lot. If P lies on or below CD we accept the lot and if P lies in between AB and CD, we continue the inspection.

▶▶ NOTE

A_m and R_m should be integers. They should be computed before inspection starts. Now if A_m is not an integer, we replace it by the largest integer less than A_m, i.e., the greatest integer contained in A_m. If R_m is not an integer, we replace it by the smallest integer greater than this.

Example 9 *Obtain the rejection and acceptance number under* SPRT *when* $\alpha = 0.08$ *and* $\beta = 0.03$, *given* $p_0 = 0.05$ *and* $p_1 = 0.10$.

Solution: Here

$$\log A = \log \frac{1-\beta}{\alpha} = \log \frac{0.97}{0.08} = 1.0837$$

$$\log B = \log \frac{\beta}{1-\alpha} = \log \frac{0.03}{0.92} = -1.4867.$$

Also

$$\log \frac{p_1}{p_0} = \log \frac{0.10}{0.05} = 0.3010; \ \log \frac{1-p_1}{1-p_0} = \log \frac{0.90}{0.95} = -0.0235.$$

Hence

$$R_m = \frac{\log A - m \log \dfrac{1-p_1}{1-p_0}}{\log \dfrac{p_1}{p_0} - \log \dfrac{1-p_1}{1-p_0}} = \frac{1.0837 + m(0.0235)}{0.3010 + 0.0235} = 3.3396 + 0.0724m$$

$$A_m = \frac{\log \frac{\beta}{1-\alpha} - m \log \frac{1-p_1}{1-p_0}}{\log \frac{p_1}{p_0} - \log \frac{1-p_1}{1-p_0}} = \frac{-1.4867 + m(0.0235)}{0.3010 + 0.0235} = -4.939 + 0.0724m.$$

Now we draw two parallel lines AB and CD (as in Fig 8) and mark the acceptance region which is the region lying below A_m inclusive and the rejection region which is the region above the line R_m inclusive and the region of continuation between the two lines. These results are presented in Table 2 below.

Table 2: Acceptance and Rejection Regions under SPRT

Number Inspected m	Acceptance Number A_m	Rejection Number R_m
0	No acceptance before	No rejection before
1	the first 68 units are	the first four units
2	inspected	are inspected
3		
4		
5		
⋮		
68		
⋮		

Example 10 *Draw the five points* OC *function to test* $H_0 : \theta = \theta_0$ *against* $H_1 : \theta = \theta_1$, *where* θ *is a Bernoullian Parameter. Consider the special case when* $\theta_0 = 0.05$, $\theta_1 = 0.10$, $\alpha = 0.08$ *and* $\beta = 0.03$.

Solution: We have seen in section 1.21.1,

$$\left(\frac{1-\theta_1}{1-\theta_0}\right)^{h(\theta)} - \theta\left(\frac{1-\theta_1}{1-\theta_0}\right)^{h(\theta)} + \theta\left(\frac{1-\theta_1}{1-\theta_0}\right)^{h(\theta)} = 1.$$

and also

$$L(\theta) = \frac{A^{h(\theta)} - 1}{A^{h(\theta)} - B^{h(\theta)}}.$$

These two equations can give the OC function by solving $h(\theta)$ from (1.21.3). Since it is often complicated to obtain value of $h(\theta)$ from (1.21.3), the five points OC function can be drawn by plotting the five easily computed points on the curve.

Thus, $L(\theta) = 1, L(\theta_0) = 1 - \alpha, L(\theta_1) = \beta$ and $L(1) = 0$.

Also there is a special point s for which

$$\begin{aligned} L(s) &= \lim_{h \to 0} \frac{A^{h(\theta)} - 1}{A^{h(\theta)} - B^{h(\theta)}} = \lim_{h \to 0} \frac{A^{h(\theta)} \log_e A}{A^{h(\theta)} \log_e A - B^{h(\theta)} \log_e B} \\ &= \frac{\log A}{\log A - \log B}; \text{ by L'Hospital's theorem.} \end{aligned} \tag{1.21.14}$$

The value of s is also obtainable by taking $h \to 0$ in the left side of (1.21.2).

$$\Rightarrow \lim_{h\to 0}\left(\frac{1-\theta_1}{1-\theta_0}\right)^{h(\theta)}\log\left(\frac{1-\theta_1}{1-\theta_0}\right)-\theta\left(\frac{1-\theta_1}{1-\theta_0}\right)^{h(\theta)}\log\left(\frac{1-\theta_1}{1-\theta_0}\right)+\theta\left(\frac{\theta}{\theta_0}\right)^{h(\theta)}\log\left(\frac{\theta_1}{\theta_0}\right)=0$$

$$\Rightarrow \log\left(\frac{1-\theta_1}{1-\theta_0}\right)-\theta\log\left(\frac{1-\theta_1}{1-\theta_0}\right)+\theta\log\left(\frac{\theta_1}{\theta_0}\right)=0$$

$$\Rightarrow \theta\left[\log\left(\frac{1-\theta_1}{1-\theta_0}\right)-\log\left(\frac{\theta_1}{\theta_0}\right)\right]=\log\left(\frac{1-\theta_1}{1-\theta_0}\right)$$

$$\Rightarrow \theta = s = \frac{\log\dfrac{1-\theta_0}{1-\theta_1}}{\log\dfrac{\theta_1}{\theta_0}-\log\dfrac{1-\theta_1}{1-\theta_0}} \tag{1.21.15}$$

and $L(s)$ is given by (1.21.14) and s in (1.21.15). Thus we have the five points OC function as given in Table 3.

►► NOTE

$\theta = \theta_0$ and $\theta = \theta_1$ are the producer's and consumer's value respectively. $\theta = 0$ and $\theta = 1$ are the boundary points.

Table 3: The Five Points OC function

θ	L(θ)
$\theta = 0.00$	1.00
$\theta_0 = 0.05$	0.92
s = 0.07	0.40
$\theta_1 = 0.10$	0.03
$\theta = 1.00$	0.00

Here, for $\theta_0 = 0.05$, $\alpha = 0.08$; $\theta_1 = 0.10$, $\beta = 0.05$.

$$A = \frac{1-\beta}{\alpha} = \frac{1-0.05}{0.08} = \frac{0.95}{0.08} = 11.875 \text{ and } B = \frac{\beta}{1-\alpha} = \frac{0.05}{0.92} = 0.054.$$

1.22 Sequential Testing for Variance when mean is known for a Normal Population

Test $H_0 : \sigma^2 = \sigma_0^2$ against $H_1 : \sigma^2 = \sigma_1^2$ for a Normal Population when the mean θ is known. We have

$$\frac{p_{1m}}{p_{0m}} = \frac{\dfrac{1}{(\sqrt{2\pi}\,\sigma_1)^m}e^{-\frac{1}{2\sigma_1^2}\sum_i(x_i-\theta)^2}}{\dfrac{1}{(\sqrt{2\pi}\,\sigma_0)^m}e^{-\frac{1}{2\sigma_0^2}\sum_i(x_i-\theta)^2}} = \left(\frac{\sigma_0}{\sigma_1}\right)^m e^{-\frac{1}{2\sigma_0^2}\sum_i(x_i-\theta)^2}\, e^{-\frac{1}{2\sigma_0^2}\sum_i(x_i-\theta)^2}$$

$$\log \frac{p_{1m}}{p_{0m}} = m \log \frac{\sigma_0}{\sigma_1} + \frac{1}{2\sigma_0^2} \sum_i (x_i - \theta)^2 - \frac{1}{2\sigma_1^2} \sum_i (x_i - \theta)^2$$

$$= m \log \frac{\sigma_0}{\sigma_1} + \sum_i (x_i - \theta)^2 \left[\frac{1}{2\sigma_0^2} - \frac{1}{2\sigma_1^2} \right].$$

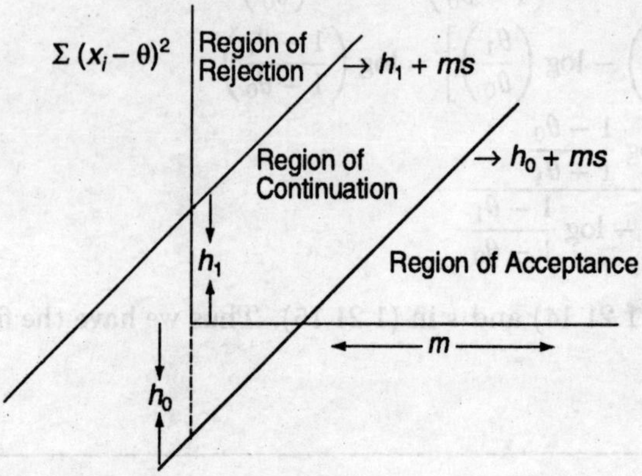

Fig 9

Hence we continue sampling when

$$\log \frac{\beta}{1-\alpha} < m \log \frac{\sigma_0}{\sigma_1} + \sum_{i=1}^{m} (x_i - \theta)^2 \left[\frac{1}{2\sigma_0^2} - \frac{1}{2\sigma_1^2} \right] < \log \frac{1-\beta}{\alpha}$$

$$\Rightarrow \quad \frac{2 \log \frac{\beta}{1-\alpha} + m \log \left(\frac{\sigma_0}{\sigma_1} \right)^2}{\frac{1}{\sigma_0^2} - \frac{1}{\sigma_1^2}} < \sum_i (x_i - \theta)^2 < \frac{2 \log \frac{1-\beta}{\alpha} + m \log \left(\frac{\sigma_0}{\sigma_1} \right)^2}{\frac{1}{\sigma_0^2} - \frac{1}{\sigma_1^2}}$$

$$\text{or,} \quad h_0 + ms < \sum (x_i - \theta)^2 < h_1 + ms. \tag{1.22.1}$$

The inequality (1.98) is exhibited in Fig 9 on page 48, where

$$h_0 = \frac{2 \log \frac{\beta}{1-\alpha}}{\frac{1}{\sigma_0^2} - \frac{1}{\sigma_1^2}}; \; h_1 = \frac{2 \log \frac{1-\beta}{\alpha}}{\frac{1}{\sigma_0^2} - \frac{1}{\sigma_1^2}} \text{ and } s = \frac{\log \frac{\sigma_1^2}{\sigma_0^2}}{\frac{1}{\sigma_0^2} - \frac{1}{\sigma_1^2}}. \tag{1.22.2}$$

1.22.1 OC function of SPRT for variance

$$L(\sigma) = \frac{A^{h(\sigma)} - 1}{A^{h(\sigma)} - B^{h(\sigma)}} = \frac{\left(\frac{1-\beta}{\alpha} \right)^{h(\sigma)} - 1}{\left(\frac{1-\beta}{\alpha} \right)^{h(\sigma)} - \left(\frac{\beta}{1-\alpha} \right)^{h(\sigma)}},$$

where $h(\sigma) = h$ is determined by the condition

$$E\left(\frac{f(x;\theta,\sigma_1)}{f(x;\theta,\sigma_0)}\right)^h = 1$$

$$\Rightarrow \quad E\left[\frac{\frac{1}{\sqrt{2\pi}\,\sigma_1}\,e^{-\frac{1}{2\sigma_1^2}\,(x-\theta)^2}}{\frac{1}{\sqrt{2\pi}\,\sigma_0}\,e^{-\frac{1}{2\sigma_1^0}\,(x-\theta)^2}}\right]^h = 1$$

$$\Rightarrow \quad \left(\frac{\sigma_0}{\sigma_1}\right)^h \frac{1}{\sqrt{2\pi}\,\sigma}\int_{-\infty}^{\infty} e^{\frac{h}{2\sigma_0^2}}\,(x-\theta)^2 - e^{-\frac{1}{2\sigma_1^2}}\,(x-\theta)^2 e^{-\frac{h}{2\sigma^2}}\,(x-\theta)^2 dx = 1$$

$$\Rightarrow \quad \left(\frac{\sigma_0}{\sigma_1}\right)^h \frac{1}{\sqrt{2\pi}\,\sigma}\int_{-\infty}^{\infty} e^{-(x-\theta)^2\left[-\frac{h}{2\sigma_0^2}+\frac{h}{2\sigma_1^2}+\frac{1}{2\sigma^2}\right]}\,dx = 1$$

$$\Rightarrow \quad \left(\frac{\sigma_0}{\sigma_1}\right)^h \frac{1}{\sqrt{2\pi}\,\sigma}\int_{-\infty}^{\infty} e^{-k\frac{(x-\theta)^2}{2}}\,dx = 1, \text{ where } k = -\frac{h}{\sigma_0^2}+\frac{h}{\sigma_1^2}+\frac{h}{\sigma^2}$$

$$\Rightarrow \quad \frac{1}{\sqrt{2\pi}\,\sigma}\left(\frac{\sigma_0}{\sigma_1}\right)^h \frac{1}{\sqrt{k}}\int_{-\infty}^{\infty} e^{-\xi^2/2}\,d\xi = 1$$

$$\Rightarrow \quad \frac{1}{\sqrt{2\pi}\,\sigma}\left(\frac{\sigma_0}{\sigma_1}\right)^h \frac{1}{\sqrt{k}}\sqrt{2\pi} = 1$$

$$\Rightarrow \quad \frac{1}{\sigma}\left(\frac{\sigma_0}{\sigma_1}\right)^h \frac{1}{\sqrt{\frac{h}{2\sigma_1^2}-\frac{h}{2\sigma_0^2}+\frac{h}{2\sigma^2}}} = 1$$

$$\Rightarrow \quad \sigma\sqrt{\frac{h}{2\sigma_1^2}-\frac{h}{2\sigma_0^2}+\frac{h}{2\sigma^2}} = \left(\frac{\sigma_0}{\sigma_1}\right)^h$$

$$\Rightarrow \quad \sigma^2\left[\frac{h}{\sigma_1^2}-\frac{h}{\sigma_0^2}+\frac{1}{\sigma^2}\right] = \left(\frac{\sigma_0}{\sigma_1}\right)^{2h}$$

$$\Rightarrow \quad \sigma^2\left[\frac{h}{\sigma_1^2}-\frac{h}{\sigma_0^2}\right] + 1 = \left(\frac{\sigma_0}{\sigma_1}\right)^{2h}$$

$$\Rightarrow \quad \sigma^2 = \frac{\left(\frac{\sigma_0}{\sigma_1}\right)^{2h} - 1}{\frac{h}{\sigma_1^2}-\frac{h}{\sigma_0^2}}$$

$$\Rightarrow \quad \sigma = \frac{\sqrt{\left(\frac{\sigma_0}{\sigma_1}\right)^{2h} - 1}}{\sqrt{\frac{h}{\sigma_1^2}-\frac{h}{\sigma_0^2}}}. \tag{1.22.3}$$

$$L(\sigma) = \frac{\left(\frac{1-\beta}{\alpha}\right)^h - 1}{\left(\frac{1-\beta}{\alpha}\right)^h - \left(\frac{\beta}{1-\alpha}\right)^h}. \tag{1.22.4}$$

Now

$$\lim_{h\to 0} \sigma = \lim_{h\to 0} \frac{\sqrt{\left(\frac{\sigma_0}{\sigma_1}\right)^{2h}\left(2\log\frac{\sigma_0}{\sigma_1}\right)}}{\sqrt{\left(\frac{1}{\sigma_0^2} - \frac{1}{\sigma_1^2}\right)}} = \frac{\sqrt{\log\left(\frac{\sigma_0}{\sigma_1}\right)^2}}{\sqrt{\left(\frac{1}{\sigma_0^2} - \frac{1}{\sigma_1^2}\right)}} = \sqrt{s}. \tag{1.22.5}$$

Further if $\sigma = 0$ $\quad L(0) = 0$

$\sigma = \sigma_0$, producer's value $\quad L(\sigma_0) = 1 - \alpha$

$\sigma = \sigma_1$, consumer's value $\quad L(\sigma_1) = 1 - \beta$

$\sigma = \infty$ $\quad L(\infty) = 0.$

Besides the four points, from (1.22.4), using L'Hospital's rule,

$$\begin{aligned}
\lim_{h\to 0} L(\sigma) &= \lim_{h\to 0} \frac{\left(\frac{1-\beta}{\alpha}\right)^h \log\left(\frac{1-\beta}{\alpha}\right)}{\left(\frac{1-\beta}{\alpha}\right)^h \log\left(\frac{1-\beta}{\alpha}\right) - \left(\frac{\beta}{1-\alpha}\right)^h \log\left(\frac{\beta}{1-\alpha}\right)} \\
&= \frac{\log\left(\frac{1-\beta}{\alpha}\right)}{\log\left(\frac{1-\beta}{\alpha}\right) - \log\left(\frac{\beta}{1-\alpha}\right)} \\
&= \frac{2\log\left(\frac{1-\beta}{\alpha}\right) \Big/ \left(\frac{1}{\sigma_0^2} - \frac{1}{\sigma_1^2}\right)}{2\log\left(\frac{1-\beta}{\alpha}\right) \Big/ \left(\frac{1}{\sigma_0^2} - \frac{1}{\sigma_1^2}\right) - 2\log\left(\frac{\beta}{1-\alpha}\right) \Big/ \left(\frac{1}{\sigma_0^2} - \frac{1}{\sigma_1^2}\right)}. \\
&= \frac{h_1}{h_1 - h_0} \text{ [from (1.22.2)].}
\end{aligned}$$

$$\therefore \lim_{h\to 0} L(\sqrt{\sigma}) \Rightarrow L(\sqrt{s}) = \frac{h_1}{h_1 - h_0}, \text{ by (1.22.5).} \tag{1.22.6}$$

1.22.2 ASN for the SPRT for testing the variance of a normal population

ASN is given by

$$\frac{E_\theta(Z_1 + Z_2 + \cdots + Z_n)}{E_\theta(Z)} = \frac{\left[\log\frac{\beta}{1-\alpha}\,(L(\sigma)) + \log\frac{1-\beta}{\alpha}\,(1 - L(\sigma))\right]}{E_\theta(Z)},$$

where

$$Z = \log \frac{f(x,\sigma_1)}{f(x,\sigma_0)} = \log \frac{\frac{1}{\sqrt{2\pi}\,\sigma_1} e^{-\frac{1}{2\sigma_1^2}(x-\theta)^2}}{\frac{1}{\sqrt{2\pi}\,\sigma_0} e^{-\frac{1}{2\sigma_0^2}(x-\theta)^2}} = \log \frac{\sigma_0}{\sigma_1} + (x-\theta)^2 \left[\frac{1}{2\sigma_0^2} - \frac{1}{2\sigma_1^2}\right]$$

$$\Rightarrow E_\theta(Z) = \log \frac{\sigma_0}{\sigma_1} + \frac{\sigma^2}{2}\left[\frac{1}{\sigma_0^2} - \frac{1}{\sigma_1^2}\right].$$

$$\text{ASN} = \frac{L(\sigma)\log\frac{\beta}{1-\alpha} + (1-L(\sigma))\log\frac{1-\beta}{\alpha}}{\frac{\sigma^2}{2}\left[\frac{1}{\sigma_0^2} - \frac{1}{\sigma_1^2}\right] + \log\frac{\sigma_0}{\sigma_1}}$$

$$= \frac{L(\sigma)\left[\left(2\log\frac{\beta}{1-\alpha}\Big/\frac{1}{\sigma_0^2} - \frac{1}{\sigma_1^2}\right) - \left(2\log\frac{1-\beta}{\alpha}\Big/\frac{1}{\sigma_0^2} - \frac{1}{\sigma_1^2}\right)\right] + h_1}{\sigma^2 + 2\log\frac{\sigma_0}{\sigma_1}\left[\frac{1}{\sigma_0^2} - \frac{1}{\sigma_1^2}\right]}.$$

$$\therefore\ \text{ASN} = \frac{L(\sigma)[h_0 - h_1] + h_1}{\sigma^2 - s}. \tag{1.22.7}$$

We have from (1.22.5)

$$\lim_{h\to 0} \sigma = \sqrt{s} \Rightarrow \lim_{h\to 0} \sigma^2 = s$$

$$\lim_{h\to 0} L(\sigma) = \frac{h_1}{h_1 - h_0}$$

$$E_\theta(n) = \frac{E_\theta(S_n)^2}{E_\theta(Z^2)} = \frac{(\log B)^2 L(\sigma) + (\log A)^2[1 - L(\sigma)]}{E_\theta(Z^2)}$$

$$E_\theta(n) = -\frac{\log A \log B}{E_\theta(Z^2)}.$$

Again

$$\lim_{h\to 0} E_\theta(z^2) = 2\left[\log\frac{\sigma_0}{\sigma_1}\right].$$

$$\text{ASN} = \frac{-\log\frac{1-\beta}{\alpha}\log\frac{\beta}{1-\alpha}}{\log\left(\frac{\sigma_0}{\sigma_1}\right)^2} \tag{1.22.8}$$

$$= \frac{-\left[2\log\frac{1-\beta}{\alpha}\Big/\left(\frac{1}{\sigma_0^2} - \frac{1}{\sigma_1^2}\right)\right]\left[2\log\frac{\beta}{1-\alpha}\Big/\left(\frac{1}{\sigma_0^2} - \frac{1}{\sigma_1^2}\right)\right]}{4\log\left(\frac{\sigma_0}{\sigma_1}\right)^2\Big/\left[\frac{1}{\sigma_0^2} - \frac{1}{\sigma_1^2}\right]^2} \tag{1.22.9}$$

$$= \frac{h_0 h_1}{2\left[\log\left(\frac{\sigma_0}{\sigma_1}\right)^2\right]^2\Big/\left[\frac{1}{\sigma_0^2} - \frac{1}{\sigma_1^2}\right]^2} = \frac{-h_0 h_1}{2\left[\log\left(\frac{\sigma_0}{\sigma_1}\right)^2\right]^2\Big/\left[\frac{1}{\sigma_1^2} - \frac{1}{\sigma_0^2}\right]^2}$$

$$= \frac{-h_0 h_1}{2s^2}\ \text{[from (1.22.2)]}. \tag{1.22.10}$$

1.22.3 Wald's identity

The procedure of obtaining ASN shown in (1.21.3) may alternatively be done by using Wald's Identity.

Let $S_n = \sum_{i=1}^{n} Z_i$, where Z_i's are independent identically distributed random variables distribution as Z. And if

- $E_\theta(Z)$ exists.
- $M(\theta) = E(e^{\theta Z})$ exists for some θ (i.e., m.g.f. exists) and $|M(\theta)| \leq 1$.
- $P(e^Z < 1-\delta) > 0$, $P(e^Z > 1+\delta) > 0$ for some $\delta > 0$, then

$$E(e^{\theta S_n}(M(\theta)^{-n}) = 1, \tag{1.22.11}$$

which is known as *Wald's Identity.*

▶▶ NOTE

In fact, $e^{\theta S_m}(M(\theta))^{-n}$ is a Martingale over $\mathcal{F}(S_n)$, where $\mathcal{F}(S_n)$ is a σ-field over S_n [*vide* Karlin, S. and Taylor, H. M. (1975)].

Determination of ASN

By differentiating Wald's Identity (1.22.11) with respect to θ and putting $\theta = 0$,

$$E\left[e^{\theta S_n} S_n (M(\theta))^{-n} - n(M(\theta)^{-n-1}(M'(\theta)e^{\theta S_n}\right] = 0,$$

we have $M(0) = 1$ and $M'(0) = E(Z)$

$$E_\theta(n) = \text{ASN} = \frac{E_\theta(S_n)}{E_\theta(Z)} \tag{1.22.12}$$

provided $E_\theta(Z) \neq 0$.

1.22.4 A comparison of ASN of SPRT with that of fixed sample size (Neyman-Pearsonian testing of hypothesis)

Let us make the comparison while considering the testing problem $H_0: \theta = \theta_0$ against $\theta = \theta_1$, where $X \sim N(\theta, \sigma^2)$, σ^2 being known.

The Neyman-Pearsonian critical region $\bar{X} > c$ is given by

$$P(\bar{X} > c \mid \theta_0) = \alpha \Rightarrow P\left(\frac{\bar{X} - \theta_0}{\sigma/\sqrt{n}} > \frac{c - \theta_0}{\sigma/\sqrt{n}} \middle| \theta_0\right) = \alpha.$$

Let us put

$$\tau_0 = \left(\frac{c - \theta_0}{\sigma/\sqrt{n}}\right).$$

The value of Type II error β is given by

$$P(\bar{X} \le c \mid \theta_1) = \beta \Rightarrow P\left(\frac{\bar{X}-\theta_1}{\sigma/\sqrt{n}} \le \frac{c-\theta_1}{\sigma/\sqrt{n}}\middle|\theta_1\right) = \beta.$$

Let us put

$$\tau_1 = \left(\frac{c-\theta_1}{\sigma/\sqrt{n}}\right) \quad \Rightarrow \quad \tau_0 - \tau_1 = \frac{c-\theta_0}{\sigma/\sqrt{n}} - \frac{c-\theta_1}{\sigma/\sqrt{n}} = \frac{\sqrt{n}\,(\theta_1-\theta_0)}{\sigma}$$

$$\Rightarrow \quad \frac{(\tau_0-\tau_1)\sigma}{(\theta_1-\theta_0)} = \sqrt{n}$$

$$\Rightarrow \quad \frac{(\tau_0-\tau_1)^2\sigma^2}{(\theta_1-\theta_0)^2} = n, \tag{1.22.13}$$

which gives the fixed sample size under NP testing procedure.

Now we consider ASN under SPRT, where $H_0 : \theta = \theta_0$.

ASN under

$$H_0 = \frac{(L(\theta_0))\log\dfrac{\beta}{1-\alpha} + (1-L(\theta_0))\log\dfrac{1-\beta}{\alpha}}{E_{\theta_0}(Z)},$$

where

$$Z = \log\frac{f(x,\theta_1)}{f(x,\theta_0)} \text{ and } E_\theta(Z) = \frac{1}{2\sigma^2}[2\theta - (\theta_1+\theta_0)](\theta_1-\theta_0).$$

Putting $\theta = \theta_0$

$$E_{\theta_0}(Z) = -\frac{1}{2\sigma^2}(\theta_1-\theta_0)^2. \tag{1.22.14}$$

$\therefore$ ASN under

$$H_0 = \frac{(1-\alpha)\log\dfrac{\beta}{1-\alpha} + \alpha\log\dfrac{1-\beta}{\alpha}}{-(\theta_1-\theta_0)^2/2\sigma^2}. \tag{1.22.15}$$

Again ASN under $H_1 : \theta = \theta_1$ is given by

$$\frac{(L(\theta_1))\log\dfrac{\beta}{1-\alpha} + (1-L(\theta_1))\log\dfrac{1-\beta}{\alpha}}{E_{\theta_1}(Z)} = \frac{\beta\log\dfrac{\beta}{1-\alpha} + (1-\beta)\log\dfrac{1-\beta}{\alpha}}{(\theta_1-\theta_0)^2/2\sigma^2}. \tag{1.22.16}$$

$$\frac{\text{ASN under } H_0}{n} = \frac{2\sigma^2\left[(1-\alpha)\log\dfrac{\beta}{1-\alpha} + \alpha\log\dfrac{1-\beta}{\alpha}\right](\theta_1-\theta_0)^2}{-(\theta_1-\theta_0)^2(\tau_0-\tau_1)^2\sigma^2}$$

$$= \frac{-2\left[(1-\alpha)\log\dfrac{\beta}{1-\alpha} + \alpha\log\dfrac{1-\beta}{\alpha}\right](\theta_1-\theta_0)^2}{(\tau_0-\tau_1)^2} \tag{1.22.17}$$

is a function of α and β only. Similarly

$$\frac{\text{ASN under } H_1}{n} = \frac{2\sigma^2\left[\beta\log\dfrac{\beta}{1-\alpha} + (1-\beta)\log\dfrac{1-\beta}{\alpha}\right](\theta_1-\theta_0)^2}{(\theta_1-\theta_0)^2(\tau_0-\tau_1)^2\sigma^2}$$

$$= \frac{2\left[\beta\log\dfrac{\beta}{1-\alpha} + (1-\beta)\log\dfrac{1-\beta}{\alpha}\right](\theta_1-\theta_0)^2}{(\tau_0-\tau_1)^2}. \tag{1.22.18}$$

It may be noted that ASN under SPRT (fixed sample size) remains independent of θ_0 and θ_1. On the average, there is approximately 33 per cent saving in the sample size by SPRT.

Example 11 *How large the sample size n must be under fixed sample and* SPRT *testing of the hypothesis* $H_0: \theta = \theta_0$ *against* $H_1: \theta = \theta_1$, *where* $X \sim N(0,1)$?

Solution: NP testing gives

$$P(\bar{X} \geq A \mid H_0) = \alpha \Rightarrow \int_A^\infty \frac{\sqrt{n}}{\sqrt{2\pi}} e^{-\frac{n}{2}(\bar{x}-\theta_0)^2}\, d\bar{x} = \alpha.$$

Putting $A = \theta_0 + \eta_\alpha/\sqrt{n}$, where $P(\tau \geq \eta_\alpha) = \alpha$, $\tau \to N(0,1)$.
Similarly, $A = \theta_1 + \eta_\beta/\sqrt{n}$, where $P(\tau \leq \eta_\beta) = \beta \Rightarrow \theta_0 + \eta_\alpha/\sqrt{n} = \theta_1 + \eta_\beta/\sqrt{n}$.
Hence

$$n = \frac{(\eta_\alpha - \eta_\beta)^2}{(\theta_1 - \theta_0)^2} \text{ and } A = \frac{\eta_\alpha\theta_1 - \eta_\beta\theta_0}{(\eta_\alpha - \eta_\beta)}.$$

In particular, if we take $\theta_0 = 0$, $\theta_1 = 1$, $\alpha = 0.05$ and $\beta = 0.05$, then $\sqrt{n} = 3.3 \Rightarrow n \cong 11$.
Again by SPRT for $H_0:\ \theta_0 = 0$ against $H_1:\ \theta_1 = 1$.

$$\begin{aligned} \text{ASN} &= \frac{-2[(1-0.05)\log(0.05/0.95) + 0.05\log(0.95/0.05)]}{(1-0)^2} \\ &= -2[0.95\log(0.05/0.95) + 0.05\log(0.95/0.05)] \\ &= -2[-1.2148 + 0.06394] = 2.30176. \end{aligned}$$

Remarks

This is an extreme case of superiority of SPRT over NP testing. In practice, the level of difference in sample size (i.e., fixed sample and ASN) needs not be so high.

Example 12 *Show that the one-sided sequential test for the mean of a Normal Population with known variance* σ^2 *may be performed by plotting the lines*

$$\begin{aligned} Y &= a\log A + cn \\ \textit{and } Y &= b\log B + cn, \end{aligned}$$

where

$$a = \frac{\sigma^2 \log A}{(\theta_1 - \theta_0)},\ b = \frac{\sigma^2 \log B}{(\theta_1 - \theta_0)} \textit{ and } c = \frac{\theta_1 + \theta_0}{2}.$$

[***Hints:*** **In** XY**-plane a decision is taken with sample size** n **when** Σy_i **crosses either of the lines.**]

Example 13 *For the Example* 12, *show that the power function of the test may be written in the form*

$$P(\theta) = \frac{1 - e^{2(c-\theta)b/\sigma^2}}{e^{2(c-\theta)a/\sigma^2} - e^{2(c-\theta)b/\sigma^2}}$$

and the ASN *may be written as*

$$E(n) = \frac{b + P(\theta(b-a)}{\theta - c}.$$

Example 14 *If p is the proportion of the defective in the submitted lot; the distribution of no. of the defective being $(0,1)$, test the hypothesis $H_0: \theta = \theta_0$ against $H_1: \theta = \theta_1$. Also show that the Acceptance Number*

$$a_n = \frac{\sigma^2}{\theta_1 - \theta_0} \log \frac{1-\beta}{\alpha} + n\frac{\theta_1 + \theta_2}{2}$$

and Rejection Number

$$r_n = \frac{\sigma^2}{\theta_1 - \theta_0} \log \frac{\beta}{1-\alpha} + n\frac{\theta_1 + \theta_2}{2}.$$

Example 15 *Employ* SPRT *to test the hypothesis $\lambda t = 1$ against $\lambda t = 2$ assuming $X \sim$ Poisson (λ, t). Using the sample mean as the basis of the test, determine the* CR *and* ASN *corresponding to $\alpha = \beta = 0.01$.*

Chapter 2

Control Charts and Other Related Topics

2.1 Introduction to Process Control

The basic idea of any production process is to maintain the standard of production while eliminating sources of inferior qualities in the manufacturing process. This is done by minimising the variability in the quality of the produced items. However, in any production process, some variation in quality cannot be avoided with the best of care. The related theories behind the possible devices of eliminating avoidable sources from time to time is known as *Control Chart Technique* evolved by Shewhart (1931).

The basic theory is that the variation in the quality or standard of the produced items can be divided into two distinct classes, viz., (*i*) variation due to assignable causes and (*ii*) random variation. The variations due to assignable causes are those sources of variation over which we have some degree of control and can be avoided by adoption of suitable reformations. *For example,* the quality in the production process due to change in the set of workers, change in the quality of raw materials or improving the efficiency of the same set of workers or machines failing due to prolonged application or use. In contrast to the above, random variations may be the result of many unknown and complex causes of variation over which we have, precisely, no control, so to *say,* even with the best care or the best devices of the production process. However, the effect of random sources of variation is usually small and the extent to which they exceed a certain limit is also probabilistically measurable. In any case, while elimination of random sources of variation is not possible, the precise objective of the Statistical Quality Control (SQC) is to eliminate the assignable causes of variation by suitable devices. The power of SQC, therefore, lies in its ability to reduce the variation due to assignable causes, which makes the diagnosis and correction of many production problems much more effective and thus brings substantial improvement in the quality of production process.

As long as the measurements of a given produced item can be seen to conform to a particular probability distribution (*say*, normal distribution) with a reasonable probability (*say*, 95% or 99%) of lying within normal limits, then the process is said to be in control. On the other hand, if because of some specific assignable causes of production, the measurements cease to conform to the same distribution and do not tend to lie within reasonable limits of random variation, then we say that there exists a lack of control in the production process and again suitable devices are adopted to check the process and to remove the assignable causes of variation so that the process can be restored back to control.

2.2 Shewhart's Control Chart

Shewhart's control chart is a simple device enabling to define more precisely the state of statistical control and judging when the process is out of control. The procedure is quite simple. A sample of a given size is taken at regular intervals and a chart is made by recording the estimates of some useful parameters of the process, *say,* the sample mean and the sample variance of the measurements every time. The latter, when taken over all the observations gives an estimate of the standard error of the sample mean. If some sample means fall outside the range specified by the mean of all sample means ± three times the standard error, one can very strongly doubt the lack of control in the production process under the assumption of central limit theorem. This is, precisely, the procedure of control chart for mean control chart for mean or $\bar{x}$. Similar procedures are applicable for control charts for standard deviation, range proportion of the defective, etc.

The process is, no doubt, too naive but sufficiently realistic for industrial applications. While much reliance is not made on precise probability of an observation deviating from a certain limit, the basic philosophy of control chart is based on long history of widespread industrial applications.

However, the statistical validity of the control chart is based on a process of *rational subgrouping* which we shall take up in the next section.

2.3 Basis for Rational Subgrouping

At the very outset, we note that control charts described above do not generally refer to individual observations but they refer to subgroups and some statistical measures estimated from these subgroups.

The procedure is to divide the observations into subgroups corresponding to produced articles during some period of time or other similarly related factors; *say,* the observations recorded as they may be produced in a day, some hours or minutes as the cases may be. Such observations are put into one subgroup. The procedure is known as *Rational Subgrouping.*

For example, let us consider the mean ($\bar{X}$)-chart. Let $\bar{x}_1, \bar{x}_2, \ldots, \bar{x}_n$ be the subgroup means. Then the idea of Shewhart's control chart is to prepare a chart of these sample means. If any of the subgroup means is found to lie outside the control limits, i.e., $\bar{\bar{x}} \pm 3\sigma/\sqrt{n}$, where $\bar{\bar{x}} =$ mean of all subgroup means, σ, the standard deviation and n is the sample size based on the total of all the subgroups, then the procedure lies in investigating the inherent cause of such phenomena.

It may be noted that one possible cause of such phenomena may be the inexperienced handling of the process by a set of newly employed hands on a particular day (for which readings are taken). There are several such systematic assignable causes affecting the quality of the production process detected by the method of rational subgrouping. However, for executing such operations, a practical knowledge relating to the technical aspects of the industry may be necessary.

In the next place, it is necessary to consider what should be the size of the subgroups. Shewhart suggested four as the ideal subgroup size. In the industrial use of the control chart, five seems to be the most common size of the subgroup. While the purpose of selecting such small subgroups is to reduce the variation within a group, there is a contrasting constraint that the distribution of the means of the subgroups should also be normally distributed. In case, the observations are taken from a non-normal universe or approximately normal population, the sample means need not be normally distributed unless we maintain a minimum subgroup size.

Judging from the two contrasting points of consideration it would be desirable to maintain a minimum size of the subgroup as four or five. Larger subgroups such as 10 or 20 are sometimes advantageous in respect that the control chart will be more sensitive to small variations in the process of average, i.e., the larger the subgroup size, the narrower is the control limits on chart for $\bar{x}$ and it is easier to detect the small variations. However, this is true only if subgroups are selected in such a way that the variations in the process average occur between and not within the subgroups. Generally speaking, larger the subgroups size, the more desirable it is to use the sample standard deviation as an estimate to the parameter σ when it is not known in the construction of $\bar{x}$ and σ-charts. Of course, the same may be valid provided the subgroup size exceeds 15.

In the introduction of control chart technique in an organisation, sometimes the subgroup size is taken as a variable. Again where the control chart technique is applied to data, already tabulated, the same way necessitates much larger subgroup than would ordinarily be selected.

No general rules may be laid down for frequency of subgroup. Each case should be decided on its own merits while considering both the cost of drawing the sample and the analysis of data. The benefit accrued out of action taken on the basis of the control chart device has to be weighed with the cost of drawing sample as well as the analysis. In the initial phase for the use of control chart for the drawing of quick conclusion it may be advantageous to take frequent samples. However, after the diagnosis of the fault and the subsequent correction of the same, the function of the control chart will be limited to the maintenance of the process control on current production. As such, it will be advisable to reduce the frequency sampling at that state.

2.4 Relationship between Rational Subgrouping and Control Chart

The information given by the control chart depends on the basis used for the selection of subgroups. A control chart may be thought of as a statistical test to determine whether the variation between subgroups is consistent with the variation within the subgroups. It is desired to determine whether or not a subgroup of measurements is statistically homogeneous (i.e., whether they appear to come from a constant system of chance causes). Subgroups should be chosen in a way that appears to give the maximum chance for the measurements to be alike and the maximum for the subgroups to differ one from the other.

This may be demonstrated in a striking way. Take a set of measurement which has been subgrouped according to the order of production and which shows definite lack of control as based on the evidence of $\bar{x}$ and R-charts. Write each measurement on a chip, put the chips in a bowl, mix them thoroughly and draw the chips one by one without replacement. Record the values within chips in the order drawn. Plot $\bar{x}$ and R-charts from these recorded values. If the mixing of the chips be properly performed, the new chart will show that the process is under control. Therefore, by mixing we have substituted the chance for the assignable cause as a basis for the differences between the subgroups. The basis of subgrouping calls for careful study with a view to obtaining the maximum amount of useful information from any control chart. The most obvious basis of the rational subgrouping is the order of production.

Two schemes involving the order of production as a basis of subgrouping are as follows:

- The first subgroup consists of all items produced as nearly as possible at one time; the next subgroup consists of all items produced as nearly as possible at a later time and so forth. *For example,* if the quality control officer makes his measurements at hourly intervals and the subgroup size is five, he may measure the last five items that were produced just before each hourly visit to the machine.

- One subgroup consists of products intended to be representative of all production over a given period of time; the next subgroup consists of products intended to be the representative of all productions approximately of the same quality. When the product accumulates to the stage of final production, the inspector may choose random samples from all products made since his last visit. If this is not practicable he may make five visits (i.e., $n = 5$) approximately equally spaced over a given production quantity or time with one measurement made during each visit. These five measurements continue one subgroup.

The first method follows the rule of selection of rational subgroups of maximising or minimising the chance for variation within a subgroup and also a maximum chance for variation between subgroups. It can be expected to give the best estimate of s that represents the ideal capabilities of a process obtainable if the assignable causes of variation between subgroups can be eliminated. Moreover, it provides a more sensitive measure of shifts in the process average; it makes the control chart a better guide to machine setting or to other actions intended to maintain a given process average. Thus the first method is more ideally suited to analysis of a process and process control.

However, if subgrouping is by the first method and a change in the process average takes place after one subgroup is taken and is corrected before the next subgroup, the change will not be reflected in the control chart. For this reason, the second method is sometimes preferred, where acceptance is to be based on the control chart. For acceptance purpose, the question to be asked before a choice between two schemes is whether or not two compensating shifts in process average are likely to occur between subgroups. If so, and if there is no other way to detect the bad product that might be produced between subgroups (*for example,* failure of a part to fit into an assembly), the second method of subgrouping is desirable in spite of other advantages of the first method. In machine operations such compensating shifts between subgroups are more likely to occur on hand-operated machines than on automatic machines.

Again where the second method of subgrouping is used, the interpretation of points out of control on R-chart is somewhat different from that in the first method. With the second method, a shift in the process average during the period covered by a subgroup may cause out of control points on the R-chart even though there has been no change in the process dispersion.

Theoretically, even when the first method is much better it may be necessary to employ the second method because of practical considerations associated with the measurements.

Finally, we may note that the order of production is not a sufficient basis for subgrouping, because the order in time or production remains a good basis for subgrouping to detect assignable cause of variation. However, there may be other assignable causes of variation that are not detected by taking subgroups in the order of production.

2.5 Statistical Basis of a Control Chart

The basis of hypothesis is that a combination of mean and standard deviation (or range) may be informative in giving an idea as to what proportion of a set of observations drawn from a specific distribution will lie within specified probabilistic limits. If the observations are drawn from a normal universe, then $\bar{x} \pm 3\sigma$ interval covers more than 99.73% of the observations. Therefore, when an observation lies outside the above limits it may be reasonably doubted that some assignable factors are affecting the production process and one should start looking for such causes. In such cases, some wastage of money and time in the way of stopping the production process for finding out such assignable sources of variation and removing them is justified.

Even if the observations are not coming from a normal universe, one can still get an idea of what proportion of observations are expected to lie within $\pm 3\sigma$ limits by using Chebychev's inequality as follows:

$$P\left(|\bar{x}-\mu| \leq \frac{t\sigma}{\sqrt{n}}\right) > 1 - \frac{1}{t^2}, \text{ for any real } t.$$

It may be seen that for $t = 3$

$$P\left(|\bar{x}-\mu| \leq \frac{t\sigma}{\sqrt{n}}\right) = 1 - \frac{1}{9} = \frac{8}{9} = 0.88,$$

i.e., whatever may be the distribution of observations, $\pm 3\sigma$ limits cover more than 88% because the limits given by Chebychev's inequality are rather narrow. On the other hand, for observations coming from normal populations $\pm 3\sigma$ limits cover more than 99.7%. This gives a statistical justification of $\pm 3\sigma$ limits for the control chart.

Remarks

The limits obtained from Chebychev's inequality as well as from the properties of normal distribution are based on true mean μ and true standard deviation σ. If we replace μ and σ by their estimates, *say*, $\bar{x}$ and s (however good, the estimates may be) then $\bar{x} \pm 3s/\sqrt{n}$, even for a normal population needs not include 99% (or any fixed stipulated level of probability as $\bar{x}$ and s are random variables). Therefore, even if all the observations lie within $\bar{x} \pm 3s/\sqrt{n}$, the process need not be necessarily in control. In fact, this is a drawback of control chart technique and we shall look into it while we discuss tolerance limits later.

2.6 Specific Control Charts

Control chart for mean is known as $\bar{X}$-*chart*.

2.6.1 $\bar{X}$-chart

- *When standards are not known:* Let $\bar{x}_1, \bar{x}_2, \ldots, \bar{x}_k$ be k subgroup means each based on a sample of size n from a normal population with mean μ and standard deviation σ. Then $\bar{x}$'s are distributed normally with mean μ and standard deviation $\sigma/\sqrt{n}$. $\mu \pm 3\sigma/\sqrt{n}$ is the interval enclosing more than 99% of the observations. But μ and σ being unknown, we require to replace them by their unbiased estimators. The unbiased estimator of μ is given by

$$\hat{\mu} = \frac{\bar{x}_1 + \bar{x}_2 + \cdots + \bar{x}_k}{k} = \bar{\bar{x}}$$

and the biased estimator of σ is given by

$$\bar{s} = \frac{s_1 + s_2 + \cdots + s_k}{k},$$

where

$$s_i^2 = \frac{1}{n}\sum_{j=1}^{n}(x_{ij}-\bar{x}_i)^2,\ i=1,2,\ldots,k.$$

Denoting $s_i = s$, let s/c_2 be unbiased for σ, since s is biased for σ, i.e., $E(s/c_2)=\sigma$.

►► NOTE

$$\begin{aligned}
\mathrm{E(s)} &= \frac{1}{\Gamma\left(\frac{n-1}{2}\right)}\int_0^\infty s e^{-\frac{ns^2}{2\sigma^2}}\left(\frac{ns^2}{2\sigma^2}\right)^{\frac{n-1}{2}} d\left(\frac{ns^2}{2\sigma^2}\right)\\
&= \frac{\left(\frac{n}{2\sigma^2}\right)^{-1/2}}{\Gamma\left(\frac{n-1}{2}\right)}\int_0^\infty \left(\frac{ns^2}{2\sigma^2}\right)^{1/2} e^{-\frac{ns^2}{2\sigma^2}}\left(\frac{ns^2}{2\sigma^2}\right)^{\frac{n-1}{2}-1} d\left(\frac{ns^2}{2\sigma^2}\right)\\
&= \frac{\left(\frac{n}{2\sigma^2}\right)^{-1/2}}{\Gamma\left(\frac{n-1}{2}\right)}\int_0^\infty e^{-\frac{ns^2}{2\sigma^2}}\left(\frac{ns^2}{2\sigma^2}\right)^{\frac{n}{2}-1} d\left(\frac{ns^2}{2\sigma^2}\right)\\
&= \frac{\left(\frac{n}{2\sigma^2}\right)^{-1/2}}{\Gamma\left(\frac{n-1}{2}\right)}\Gamma\left(\frac{n}{2}\right) = \frac{\sigma\sqrt{2/n}}{\Gamma\left(\frac{n-1}{2}\right)}\Gamma\frac{n}{2}\\
\Rightarrow \mathrm{E(s)} &= c_2\sigma,\ \mathrm{var\,(s)} = \sigma^2\left(\frac{n-1}{n} - c_2^2\right),\\
\text{where } c_2 &= \frac{\Gamma(n/2)}{\Gamma\left(\frac{n-1}{2}\right)}\sqrt{2/n}.
\end{aligned}$$

Hence, $\mu \pm 3\sigma/\sqrt{n}$ is estimated as

$$\bar{\bar{X}} \pm 3\frac{\bar{s}}{c_2}\frac{1}{\sqrt{n}}.$$

∴ the control chart is specified by:

$$\begin{aligned}
\text{Central line (CL):}&\quad \bar{\bar{x}}\\
\text{Upper Control Limit (UCL):}&\quad \bar{\bar{x}} + 3\bar{s}/c_2(1/\sqrt{n})\\
\text{Lower Control Limit (LCL):}&\quad \bar{\bar{x}} - 3\bar{s}/c_2(1/\sqrt{n}).
\end{aligned}$$

- *When standards are known:* We have $E(\bar{x}) = \mu$ and $\sqrt{\mathrm{var}\,(\bar{x})} = \sigma/\sqrt{n}$.

∴ if the standards are known and given by $\mu = \mu_0$ and $\sigma = \sigma_0$, then replacing μ and σ by μ_0 and σ_0, we have

$$\begin{aligned}
\text{Central line (CL):}&\quad \mu_0\\
\text{Upper Control Limit (UCL):}&\quad \mu_0 + 3\sigma_0/\sqrt{n}\\
\text{Lower Control Limit (LCL):}&\quad \mu_0 - 3\sigma_0/\sqrt{n}.
\end{aligned}$$

2.6.2 Control chart for standard deviation (S-chart)

- *When standards are not known:* We have,

$$E(s) = c_2\sigma \text{ and } \text{var}(s) = \sigma^2.\left(\frac{n-1}{n}\right) - c_2^2,$$

where c_2 is an adjusting constant and its value is given in SQC table. If σ is not known, we replace σ by $\bar{s}/c_2$ and the upper control limit is given by the estimate of $E(s)+3\sqrt{\text{var}(s)}$.

$$\begin{aligned} \text{i.e., UCL} &= c_2(\bar{s}/c_2) + 3(\bar{s}/c_2)\sqrt{\frac{n-1}{n} - c_2^2} \\ &= \bar{s} + 3(\bar{s}/c_2)\sqrt{\frac{n-1}{n} - c_2^2}. \end{aligned}$$

Similarly, the central line is estimated by

$$c_2(\bar{s}/c_2) = \bar{s} \text{ and LCL} = \bar{s} - 3(\bar{s}/c_2)\sqrt{\frac{n-1}{n} - c_2^2}.$$

- *When standards are known:* In this case, replacing σ by its known standard value, *say*, σ_0, we have

$$\begin{aligned} \text{UCL} &= c_2\sigma_0 + 3\sigma_0\sqrt{\frac{n-1}{n} - c_2^2} \\ \text{CL} &= c_2\sigma_0 \\ \text{and LCL} &= c_2\sigma_0 - 3\sigma_0\sqrt{\frac{n-1}{n} - c_2^2}. \end{aligned}$$

2.6.3 Control chart for range (R-chart)

- *When standards are not known:* Denoting the range by R and its mean over all the subgroups as $\bar{R}$, we have

$$\text{UCL} = E(R_i) + 3\sqrt{\text{var}(R_i)},$$

where R_i is the range corresponding to the ith subgroup ($i = 1, 2, \ldots, k$).
We have, $E(R_i) = d_2\sigma$ and $\text{var}(R_i) = D^2\sigma^2$, where d_2 and D are adjusting constants whose values are given in SQC tables.
Replacing σ by its estimate $\bar{R}/d_2$ we have the estimated

$$\begin{aligned} \text{UCL} &= d_2(\bar{R}/d_2) + 3D(\bar{R}/d_2) \\ &= \bar{R} + 3D(\bar{R}/d_2) \\ \text{CL} &= \bar{R} \\ \text{LCL} &= \bar{R} - 3D(\bar{R}/d_2). \end{aligned}$$

- *When standards are known:* Let σ_0 be the known standard value of σ. Then putting $\sigma = \sigma_0$ in $E(R_i)$ and $\text{var}(R_i)$, we get

$$\begin{aligned} \text{UCL} &= d_2\sigma_0 + 3D\sigma_0 \\ \text{CL} &= d_2\sigma_0 \\ \text{and LCL} &= d_2\sigma_0 - 3D\sigma_0. \end{aligned}$$

▶▶ NOTE

In case, LCL turns out to be negative it is conventionally taken as zero as the range R can never become negative.

2.6.4 Control chart for the number of defectives (np-chart)

We understand by the term defective as concerning a manufacturing item which fails to meet one or more of the specifications needed in the finally manufactured product. A particular defect is an example to show the lack of conformity of the item to meet the specifications.

If n be the subgroup size and p is the proportion of the defective, in the sample then $E(np) = nP$ and $\text{var}\,(np) = nP(1-P)$, where P is the proportion of the defective, in the population.

- *When standard is not known:* Replacing p by the subgroup mean $\bar{p}$, where $\bar{p} = \sum_{i=1}^{k} p_i/k$ and p_i being the proportion of the defective in the ith subgroup $(i = 1, 2, \ldots, k)$ we have the estimated

$$\begin{aligned} \text{UCL} &= n\bar{p} + 3\sqrt{n\bar{p}(1-\bar{p})} \\ \text{CL} &= n\bar{p} \\ \text{LCL} &= n\bar{p} - 3\sqrt{n\bar{p}(1-\bar{p})}. \end{aligned}$$

- *When standard is known:* Suppose, the known standard is given by $p = p_0$, then

$$\begin{aligned} \text{UCL} &= np_0 + 3\sqrt{np_0(1-p_0)} \\ \text{CL} &= np_0 \\ \text{LCL} &= np_0 - 3\sqrt{np_0(1-p_0)}. \end{aligned}$$

2.6.5 Control chart for proportion of defectives (p-chart)

If p is the proportion of the defective in any subgroup, then

$$E(p) = P \text{ and } \text{var}\,(p) = P(1-P)/n,$$

where P is the true proportion of the defectives.

- *When standard is not known:* We may replace p by $\bar{p}$, the sample mean of the proportion of the defectives over all subgroups and the estimated control limits are given by

$$\begin{aligned} \text{UCL} &= \bar{p} + 3\sqrt{\bar{p}(1-\bar{p})/n} \\ \text{CL} &= \bar{p} \\ \text{and LCL} &= \bar{p} - 3\sqrt{\bar{p}(1-\bar{p})/n}. \end{aligned}$$

Here also we have to take into account the event of LCL being negative just like R-chart. Since theoretically p cannot be negative, in the event of such estimated LCL, one has to conventionally take the same as zero.

Again in both the cases of p-chart and np-chart with varying control limits one may, instead of plotting p, use the standardized variate, viz.,

$$\frac{p-\bar{p}}{\sqrt{\bar{p}(1-\bar{p})/n}},$$

where $\bar{p}$ is the mean proportion over all subgroups. In this case, we have UCL = 3, LCL = −3 and CL = 0.

- *When standard is known:* If the standard is given by $p = p_0$, then replacing P by p_0 we have

$$\begin{aligned} \text{UCL} &= p_0 + 3\sqrt{p_0(1-p_0)/n} \\ \text{CL} &= p_0 \\ \text{and LCL} &= p_0 - 3\sqrt{p_0(1-p_0)/n}. \end{aligned}$$

2.6.6 Control chart for the number of defects (C-chart)

While considering the distribution of number of defects in the manufacturing process one may note two essential points, viz.,

(*a*) the scope of a defect in any position of the manufactured item is unlimited but

(*b*) the scope of a defect occurring in any particular position is very small.

These assumptions are equivalent to $n \to \infty$, $p \to 0$ subject to np is a constant which leads the distribution of the number of defects as Poisson variables. Denoting by d, the number of defects in an item we have

$$E(d) = \lambda \text{ and } \sqrt{\text{var}\,(d)} = \lambda.$$

- *When standard is not known:* In this case we may replace λ by $\bar{d} = \Sigma d_i/k$, where d_i represents the number of defects in the ith subgroup $(i = 1, 2, \ldots, k)$. Accordingly, the control limits are estimated by

$$\begin{aligned} \text{UCL} &= \bar{d} + 3\sqrt{\bar{d}} \\ \text{CL} &= \bar{d} \\ \text{LCL} &= \bar{d} - 3\sqrt{\bar{d}}. \end{aligned}$$

- *When standard is known:* Let the standard be given by $\lambda = \lambda_0$, then replacing λ by λ_0, we have

$$\begin{aligned} \text{UCL} &= \lambda_0 + 3\sqrt{\lambda_0} \\ \text{CL} &= \lambda_0 \\ \text{LCL} &= \lambda_0 - 3\sqrt{\lambda_0}. \end{aligned}$$

2.6.7 X̄-chart using R (range) instead of σ (standard deviation)

- *When standards are not known:* In this case, the control limits will be as usual estimated by

$$\begin{aligned} \text{UCL} &= \bar{\bar{x}} + 3\bar{R}/d_2\sqrt{n} \\ \text{CL} &= \bar{\bar{x}} \\ \text{LCL} &= \bar{\bar{x}} - 3\bar{R}/d_2\sqrt{n}. \end{aligned}$$

Following the same argument as in 2.5.1 and $\bar{R}$ and d_2 are same as defined in 2.5.3.

- *When standards are known:* If the standards are given by $\mu = \mu_0$ and $R = R_0$, then replacing μ and R by μ_0 and R_0 respectively, we have

$$\begin{aligned} \text{UCL} &= \mu_0 + 3R_0/\sqrt{n} \\ \text{CL} &= \mu_0 \\ \text{LCL} &= \mu_0 - 3R_0/\sqrt{n}. \end{aligned}$$

2.7 OC Function and Average Run Length (ARL) of the X̄-chart

If for $\bar{X}$-chart we use LCL as $\bar{\bar{x}} - 3.09\sigma/\sqrt{n}$ and UCL as $\bar{\bar{x}} + 3.09\sigma/\sqrt{n}$ when σ is known, then we accept all observations lying between LCL and UCL. Suppose, we use standardized variable, viz.,

$$\frac{\bar{\bar{x}} - \mu}{\sigma/\sqrt{n}} = \frac{-\theta\sqrt{n}}{\sigma},$$

where $\mu - \bar{\bar{x}} = \theta$, the deviation of the process mean from the target. Then

$$\begin{aligned} \text{LCL} &= \frac{\bar{\bar{x}} - 3.09\sigma/\sqrt{n} - \mu}{\sigma/\sqrt{n}} = -3.09 - \frac{\theta\sqrt{n}}{\sigma} \\ \text{and UCL} &= \frac{\bar{\bar{x}} + 3.09\sigma/\sqrt{n} - \mu}{\sigma/\sqrt{n}} = 3.09 - \frac{\theta\sqrt{n}}{\sigma}. \end{aligned}$$

Then the OC function of the $\bar{X}$-chart is given by

$$P(\theta) = \Phi\left(3.09 - \frac{\theta\sqrt{n}}{\sigma}\right) - \Phi\left(-3.09 - \frac{\theta\sqrt{n}}{\sigma}\right), \tag{2.7.1}$$

where

$$\Phi(t) = \frac{1}{\sqrt{2\pi}} \int_{-\infty}^{t} e^{-z^2/2}\, dz.$$

Suppose, we are making batch inspection which accepts each batch independently of others with an OC curve $P(\theta)$. Then the probability that a run length of r batches is accepted is $P(\theta)^{r-1}[1 - P(\theta)]$. Therefore, the Average Run Length (ARL) is given by

$$\begin{aligned} \text{ARL} &= 1[1 - P(\theta)] + 2P(\theta)[1 - P(\theta)] + 3[P(\theta)]^2[1 - P(\theta)] + \ldots \\ &= [1 - P(\theta)][1 + 2P(\theta) + 3[P(\theta)]^2 + \ldots] \\ &= \frac{1 - P(\theta)}{[1 - P(\theta)]^2} = \frac{1}{1 - P(\theta)} \\ &= \frac{1}{1 - \left\{\Phi\left(3.09 - \frac{\theta\sqrt{n}}{\sigma}\right) - \Phi\left(-3.09 - \frac{\theta\sqrt{n}}{\sigma}\right)\right\}}. \end{aligned} \tag{2.7.2}$$

From (2.6.1) and (2.6.2) it follows that ARL gives a measure of the extent of product as a whole on the average to be examined before a change in the mean is detected.

Remarks

The ARL function of the above $\bar{X}$-chart is rather flat. It can be sharpened if (*i*) warning limits are drawn at $\pm 1.96\sigma/\sqrt{n}$ and appropriate action may be drawn if two consecutive points fall outside the same range; (*ii*) action should be taken if a specified number, *say*, k, out of the last m points falls between the action and warning limits; (*iii*) one may also draw another set of lines at $\pm\sigma/\sqrt{n}$ and action should be taken if three consecutive points fall outside the same.

So far we have assumed that σ is a known constant. However, changes in the process mean as well as variance may indicate that the process is out of control. In that case (2.6.1) may be regarded as a function of both θ and σ. Let us consider the behaviour of ARL of $\bar{X}$-chart when both θ and σ vary. Let us assume that action is taken only when $\pm 3.09\sigma/\sqrt{n}$ is crossed. We assume that the distribution of the measurement is normal with mean θ and variance σ^2 instead of σ^2 and the ARL contours are drawn under different levels of $\theta\sqrt{n}/\sigma$ (both positive and negative) and the ratio of σ'/σ. The contours are graphically drawn in the Fig 10.

It may be seen with slight change in the ratio of σ'/σ that ARL is changed considerably. It may also be noticed that even when σ'/σ is slightly increased from one ARL changes from 100 to 50. When it is changed from one to two, ARL changes from 50 to 10 showing the high sensitivity of σ'/σ when θ undergoes slight changes. It, therefore, follows from Fig 10 that $\bar{X}$-chart alone gives very little protection against variation in σ. This shows the imperativeness of σ-chart or R-chart along with $\bar{X}$-chart before proper functioning of the latter.

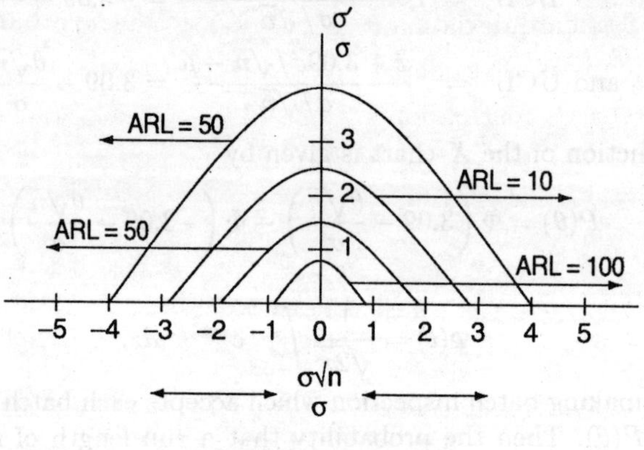

Fig 10 ARL Contours

Finally, it is important to note the rationale of applying OC function or ARL function in the control charts. They enable us to find out the size of the batch (or lot) as well as to find out the interval between taking observations of two groups. The ARL for a unit time interval has been shown to be $1/[1 - P(\theta, \sigma, n)]$ when θ, σ and n all vary. Therefore, ARL for a time interval is $t/[1 - P(\theta, \sigma, n)]$.

If we put an upper limit of ARL for a given value of θ as u_θ, then

$$u_\theta = t/[1 - P(\theta, \sigma, n)] \Rightarrow P(\theta, \sigma, n) = (1 - t/u_\theta).$$

But $P(\theta, \sigma, n)$ is a function of $\theta\sqrt{n}/\sigma$ and a graph of t/μ_θ for values of $\theta\sqrt{n}/\sigma$ would essentially be the OC function. However, for such a graph it is possible to obtain a series of values of t and n. A final selection can be made on the basis of ARL when $\theta = 0$.

2.8 A Special Advantage of R-chart

It may be noted that the OC or ARL function of the R-chart is independent of the process mean μ and depends only on σ. Therefore, it is easier to handle them than the $\bar{X}$-chart which may depend on both the process mean and the variance. If σ' is the true variance, then an advantageous property of the distribution of R/σ' is that the same is independent of σ'. Therefore, if R-chart is drawn with UCL $= R_U$ and LCL $= R_L$, then all that we should do is to find out the probabilities, viz.,

$$P(R/\sigma'/ > R_U) \text{ and } P(R/\sigma' < R_L).$$

This enables us to draw the ARL curve of R-chart.

Example 1 *Desmond* (1954) *presents a problem dealing with inspection of voltage regulations for private motor cars. It is assumed that the voltage regulators operate in the range of* 15.8 *to* 16.4 volts. *However, the experiment revealed a very high level of measurement error with a variance of* 0.0511. *The error of measurement, therefore, had to be superimposed on the true natural variation of the true regulator operating voltage whose variance was found to be of the order of* 0.0435. *Under this set up what would be the corrected Shewhart's control limits at* 99%?

Solution:

$$\begin{aligned}
\text{The total variance} &= 0.0511 + 0.0435 = 0.0946\\
\text{Standard deviation } \hat{\sigma} &= \sqrt{0.0946} = 0.308\\
\text{Average operation level of regulators} &= 1/2(15.8 + 16.4) = 16.1\\
\therefore \text{ Shewhart's control limits} &= 16.1 \pm 3.09 \times 0.308\\
&= 15.15, 17.05 \text{ volts}
\end{aligned}$$

$\Rightarrow$ LCL $= 15.15$ and UCL $= 17.05$.

However, in case it is shown that the production is in control, we can use 99.8% limits as

$$\begin{aligned}
\text{UCL} &= 16.1 + 3.09 \times 0.209 = 16.75\\
\text{and LCL} &= 16.1 - 3.09 \times 0.209 = 15.45 \ (\because \ \sqrt{0.0435} = 0.209).
\end{aligned}$$

2.9 Economic Design of X̄-chart

We are basically confronted with the question, viz., how a user of the control chart should choose the design parameter. As a matter of fact, the design of a control chart has several economic consequences that can be expressed in terms of different types of costs. In this connection, the following types of costs are worthwhile to consider, viz.,

(*i*) Costs of sampling and testing.

(*ii*) Costs associated with the detection of faults by inspection of units in each sample following investigation of *out of control* signals and finally correcting assignable causes if any.

(*iii*) Costs of transmitting non-conformable units (it may be noted that any non-conforming item need not necessarily be defective) to reach the consumer.

Without going into details in respect of the exact possible categories of costs under items (*i*), (*ii*) and (*iii*) given above, it may be possible to incorporate the cost factors in the design parameters of control charts. Therefore, logically it is more worthwhile to consider the design of a control chart from an economic point of view. Duncan (1956) was the pioneer to consider the issue. He recommended replacing general guidelines with a process specific design. The basic motivation in such economic designs is to minimize the net sum of all types of costs involved in the process.

In the next section of the study we present a survey of economic control chart ($\bar{X}$-chart) under single assignable cause, responsible for lack of control in the system. The general approach for such economic designs is to build up a cost model while taking into consideration of such components or factors which should come in the way of calculating the costs. This enables us to construct a cost function. The cost function, so formulated, then is required to be optimized with respect to the design parameters of the control chart.

As stated earlier, the cost models for an economic control chart was first proposed by Duncan (1956) followed by Gibra (1971) and Lorenzen and Vance (1986). Duncan has built up a model leading to minimize the total expected cost. Gibra, however, developed a model envisaging to minimize the expected cost over production cycles converted per unit time basis.

2.10 Duncan's Cost Model for the Economic Design of $\bar{X}$-chart

We describe the process generating the cost model as follows:

We draw a sample of size n from a process every h hours and plot an $\bar{X}$-chart. Given that the process is in control, the distribution of $\bar{X}$ is $N(0, \sigma^2/n)$. But we assume if the process is out of control, then $\bar{X}$ is distributed normally with the same variance but the mean changing to either $+\delta\sigma$ or $-\delta\sigma$. On the control chart we set up the following control limits, viz.,

$$\begin{aligned} \text{Central line} &= 0 \\ \text{Lower Control Limit (LCL)} &= -k\sigma/\sqrt{n} \\ \text{Upper Control Limit (UCL)} &= k\sigma/\sqrt{n} \end{aligned}$$

A search for an assignable cause is made only if a point falls outside the above control limits. Our problem lies in choosing the parameters (h, k, n) to minimize the cost (or to maximize the profit). It is assumed that the process in control initially and the time for the arrival of an assignable cause changing the mean to $\delta\sigma$ or $-\delta\sigma$ (with equal probability) is distributed exponentially with parameter λ. This gives the average time for an assignable cause as $1/\lambda$. The time taken to plot the points in the control chart is taken as proportional to n, *say gn*. Also, the average time to check and eliminate an assignable cause is denoted by T. Now if an assignable cause occurs within an interval $(0, h)$, the expected time when it occurs is given by

$$\begin{aligned}
\frac{\int_0^h t\lambda e^{-\lambda t}\,dt}{\int_0^h \lambda e^{-\lambda t}\,dt} &= \frac{1-(1+\lambda h)e^{-\lambda h}}{\lambda(1-e^{-\lambda h})} = \frac{1-(1+\lambda h)\left(1-\lambda h+\frac{\lambda^2h^2}{2}-\frac{\lambda^3h^3}{6}+\dots\right)}{\lambda\left[1-\left(1-\lambda h+\frac{\lambda^2h^2}{2}-\frac{\lambda^3h^3}{6}+\dots\right)\right]} \\
&= \frac{\frac{\lambda^2h^2}{2}-\frac{\lambda^3h^3}{3}+\frac{\lambda^4h^4}{8}-\dots}{\lambda^2h\left[\left(1-\frac{\lambda h}{2}+\frac{\lambda^2h^2}{6}-\dots\right)\right]} \\
&\cong \left[\frac{h}{2}-\frac{\lambda h^2}{3}+\frac{\lambda^2h^3}{8}-\dots\right]\left[1+\left(\frac{\lambda h}{2}-\frac{\lambda^2h^2}{6}\right)+\left(\frac{\lambda h}{2}-\frac{\lambda^2h^2}{6}\right)+\dots\right] \\
&= \frac{h}{2}-\frac{\lambda h^2}{12}+0(\lambda^3h^4)
\end{aligned}$$

and the approximation is valid for small λ and moderate h. Next, we introduce two types of errors, viz., E_I and E_{II}. A type I error may arise if there occurs a signal of an assignable cause, but really the process is under control. In other words, E_I corresponds to a false alarm signal. Type II error occurs when there is no signal for a real out of control process. Let α and β be the probabilities of E_I and E_{II} respectively. When the process goes out of control the observations are assumed to be independent identically distributed normal with mean $\pm\delta\sigma$.

$$\begin{aligned}
\beta &= P\,(\text{no sample mean lies outside the control limits} \mid \text{true mean} = \pm\delta\sigma) \\
&= P\left(\frac{-k\sigma/\sqrt{n}\mp\delta\sigma}{\sigma/\sqrt{n}} \le \frac{\bar{x}\mp\delta\sigma}{\sigma/\sqrt{n}} \le \frac{k\sigma/\sqrt{n}\mp\delta\sigma}{\sigma/\sqrt{n}}\right) \\
&= P(\mp\delta\sqrt{n}-k \le Z \le \mp\delta\sqrt{n}+k), \text{ where } Z \text{ is a standard normal variate.} \\
&= \Phi(\mp\delta\sqrt{n}+k)-\Phi(\mp\delta\sqrt{n}-k). \qquad (2.10.1)
\end{aligned}$$

Similarly, power of an $\bar{X}$-chart is given by
P (a signal for out of control state on the $\bar{X}$-chart | process is out of control)

$$= 1-\beta = [1-[\Phi(\mp\delta\sqrt{n}+k)] + \Phi[\mp\delta\sqrt{n}-k)]$$

$$\therefore\ P(k,n,\delta) = \Phi(\pm\delta\sqrt{n}+k)-\Phi(\mp\delta\sqrt{n}-k) \qquad (2.10.2)$$

which is the probability of detecting an assignable cause. Therefore, the probability of a type I error given by α is

$$\begin{aligned}
&P\,(\text{a sample point falls outside the control limit} \mid \text{true mean} = 0) \\
&= P\left(\frac{-k\sigma/\sqrt{n}\mp\delta\sigma}{\sigma/\sqrt{n}} \le \frac{\bar{x}\mp\delta\sigma}{\sigma/\sqrt{n}} \le \frac{k\sigma/\sqrt{n}\mp\delta\sigma}{\sigma/\sqrt{n}}\right)^c,
\end{aligned}$$

where c represents the complimentary set of the event, given in the bracket.

$$\begin{aligned}
&= P(-k \le Z \le k)^c, \text{ where } Z = \frac{\bar{x}-0}{\sigma/\sqrt{n}} \\
&= P(Z \le -k) + P(Z \ge k) \\
&= 2P(Z \le -k) \qquad (2.10.3)
\end{aligned}$$

because of the symmetry of the normal distribution. Therefore, from (2.10.1) and (2.10.2) it follows that

$$P(k, n, 0) = \alpha. \tag{2.10.4}$$

Now $p = P(k, n, \delta)$ is the probability of detecting an assignable cause. Suppose, the assignable cause is detected in the rth sample; $r = 1, 2, \ldots$ The probability of the same is $p(1-p)^{r-1}$. This being a Geometric distribution $E(r) = 1/p$ = mean number of samples drawn when an assignable cause occurs. Now for each sample, we have a duration of h hours. Therefore, to get an assignable cause, the waiting time on the average is h/p hours.

Next, we define the average cycle length as

C = average time in control + average time out of control till detection + plotting delay + delay in checking.

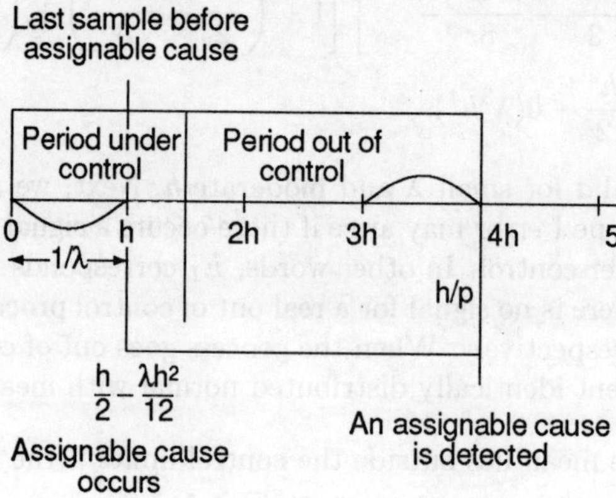

Fig 11

$$\text{Period out of control} = \frac{h}{p} - \frac{h}{2} + \frac{\lambda h^2}{12}$$

$$\Rightarrow C = \frac{1}{\lambda} + \left(\frac{h}{p} - \frac{h}{2} + \frac{\lambda h^2}{12}\right) + gn + T = \text{average cycle length,} \tag{2.10.5}$$

where gn represents the average delay in plotting n points and T represents the delay in checking and eliminating an assignable cause. Therefore, the proportion of time, the process is in control, is

$$\pi_1 = \frac{1}{C\lambda} \tag{2.10.6}$$

and the proportion of time, the process is not in control, is

$$\pi_2 = 1 - \frac{1}{C\lambda} = 1 - \pi_1. \tag{2.10.7}$$

$\therefore$ expected number of false alarms per hour

$$= \frac{\text{Expected \# of false alarms when process is in control}}{\text{average cycle length}}$$

$$= \frac{\text{Expected \# of samples during the process in control} \times P \text{ (a false alarm)}}{C}$$

$$= s\alpha/C, \tag{2.10.8}$$

where s is the expected number of samples during the process in control.

We can write,

$$s = \sum_{i=0}^{\infty} i\, P\,[\text{an assignable cause occurs in between } i\text{th and } (i+1)\text{th sample}],$$

i.e., the process was in control till ith sample ($i = 0, 1, 2, \dots$).

If ν is a random variable representing the waiting time for an assignable cause P [an assignable cause occurs between ith and $(i+1)$th sample]

$$P[ih \le \nu \le (i+1)h].$$

$$\therefore\; s = \sum_{i=0}^{\infty} iP[ih \le \nu \le (i+1)h].$$

$$\begin{aligned}
\therefore\; s &= \sum_{i=0}^{\infty} i[F_\nu(i+1)h - F_\nu(ih)] \\
&\qquad \text{Since } F_\nu(i+1)h = P[-\infty, (i+1)h] \text{ and } F_\nu(ih) = P[-\infty, (ih)] \\
&\qquad F_\nu(i+1)h - F_\nu(ih) = P[ih \le \nu \le (i+1)h] \\
&= \sum_{i=0}^{\infty} i\left[\left(1 - e^{-\lambda(i+1)h}\right) - \left(1 - e^{-\lambda ih}\right)\right] \\
&= \sum_{i=0}^{\infty} i\left(e^{-\lambda ih} - e^{-\lambda(i+1)h}\right) \\
&= \sum_{i=0}^{\infty} ie^{-\lambda ih}\left(1 - e^{-\lambda h}\right) \\
&= \left(1 - e^{-\lambda h}\right)\sum_{i=0}^{\infty} ie^{-\lambda ih} \\
&= \left(1 - e^{-\lambda h}\right)\left(e^{-\lambda h} + 2e^{-2\lambda h} + 3e^{-3\lambda h} + \dots\right) \\
&= \left(1 - e^{-\lambda h}\right)e^{-\lambda h}\left(1 + 2e^{-\lambda h} + 3e^{-2\lambda h} + \dots\right) \\
&= \frac{(1 - e^{-\lambda h})e^{-\lambda h}}{(1 - e^{-\lambda h})^2} = \frac{e^{-\lambda h}}{(1 - e^{-\lambda h})}.
\end{aligned} \tag{2.10.9}$$

Putting (2.9.9) in (2.9.8),

$$\begin{aligned}
\text{Expected \# of false alarms per hour} &= \frac{s\alpha}{C} = \frac{\alpha e^{-\lambda h}}{(1 - e^{-\lambda h})C} = \frac{\alpha(1 - \lambda h)}{(1 - (1 - \lambda h))C} && (2.10.10) \\
&\cong \frac{\alpha}{\lambda hC} - \frac{\alpha\lambda h}{\lambda hC} = \frac{\pi_1\alpha}{h} - \frac{\alpha}{C} \\
&= \frac{\pi_1\alpha}{h} - \frac{\alpha}{1/\pi_1\lambda}\ [\text{from (2.9.6)}] \\
&= \frac{\pi_1\alpha}{h} - \alpha\pi_1\lambda \cong \frac{\pi_1\alpha}{h} && (2.10.11)
\end{aligned}$$

since $\alpha\pi_1\lambda$ is a small quantity of negligible order (all the three being in proportions), (2.10.11) gives us the average loss per unit time as

$$
\begin{aligned}
L &= \text{loss when the process is out of control} + \text{cost of operating the } \bar{X}\text{-chart} + \\
&\qquad \text{cost of an assignable cause} + \text{cost of false alarms.} \\
&= \text{I} + \text{II} + \text{III} + \text{IV} \\
\text{I} &= \text{Loss when the process is out of control} \\
&= (\text{loss in profit when the process is out of control}) \times \text{probability (or proportion} \\
&\qquad \text{of time) the process is out of control} \\
&= M\pi_2,
\end{aligned}
$$

where M represents the loss in profit when the process is out of control and π_2 represents the probability of the process getting out of control.

$$
\begin{aligned}
\text{II} &= \text{Cost of operating the } \bar{X}\text{-chart per unit time} \\
&= \text{Cost of plotting the } \bar{X}\text{-chart per sample } \times 1/h = b/h \\
\text{III} &= \text{Cost of an assignable cause per hour} \\
&= \text{Cost of measuring per item} \times \text{sample size } \times 1/h + \text{cost of checking and} \\
&\qquad \text{eliminating an assignable cause per unit time.} \\
&= Cn/h + W/C, \textit{ say, } \text{where } C \equiv \text{length of the cycle} \\
\text{IV} &= \text{Cost of false alarms} \\
&= \text{Expected \# false alarms per hour} \times \text{cost of searching for an assignable cause} \\
&\qquad \text{when none exists} \\
&= \frac{\pi_1 \alpha}{h} C', \textit{ (say)},
\end{aligned}
$$

where C' is the cost of searching an assignable cause when none exists.

$$
\therefore\ L = M\pi_2 + \left(\frac{b}{h} + \frac{Cn}{h}\right) + \frac{W}{C} + C'\left(\pi_1 \frac{\alpha}{h}\right), \tag{2.10.12}
$$

where W is the cost of checking and eliminating the assignable cause.

Now $\pi_1 = \dfrac{1}{C\lambda}$;

$$
\text{we put } B = \frac{C\lambda - 1}{\lambda} = \left(\frac{1}{p} - \frac{1}{2} + \frac{\lambda h}{12}\right) h + gn + T \text{ [from (2.10.5)]}
$$

where T is the cost of searching for an assignable cause when none exists since

$$
\begin{aligned}
C &= \frac{1}{\lambda} + \left(\frac{1}{p} - \frac{1}{2} + \frac{\lambda h}{12}\right) h \Rightarrow \frac{C\lambda - 1}{\lambda} = \left(\frac{1}{p} - \frac{1}{2} + \frac{\lambda h}{12}\right) h \\
\Rightarrow M\pi_2 &= M\left(1 - \frac{1}{C\lambda}\right) = M\frac{C\lambda - 1}{C\lambda} = \frac{MB}{B + 1/\lambda} \text{ (since } C = B + 1/\lambda) \\
&= \frac{MB\lambda}{B\lambda + 1} \\
\text{and } \frac{W}{C} &= \frac{W}{B + 1/\lambda} = \frac{W\lambda}{B\lambda + 1}.
\end{aligned}
$$

Further,

$$
C'\frac{\pi_1 \alpha}{h} = \frac{C'}{C\lambda}\frac{\alpha}{h} = \frac{C'\alpha}{h}\frac{1}{(B + 1/\lambda)\lambda} = \frac{C'\alpha}{h(B\lambda + 1)}.
$$

Therefore, we can write (2.10.12) as

$$\begin{aligned} L &= \frac{MB\lambda}{B\lambda+1} + \frac{W\lambda}{B\lambda+1} + \frac{C'\alpha}{h(B\lambda+1)} + \frac{b+Cn}{h} \\ &= \frac{MB\lambda + W\lambda + C'\alpha/h}{B\lambda+1} + \frac{b+Cn}{h}. \end{aligned} \tag{2.10.13}$$

If we put $W = C'/2$; and if α and $\lambda \to 0$

$$L = \frac{MB\lambda + (C'/2)\lambda + C'\alpha/h}{B\lambda+1} + \frac{b+Cn}{h},$$

where

$$B = \left(\frac{1}{p(\lambda,k,n)} - \frac{1}{2} + \frac{\lambda h}{12}\right)h + gn + T. \tag{2.10.14}$$

Now L can be minimized with respect to n, k and h. We can minimize L for choice of h, k and n. However, in order that the results are expressed in elegant form, it is necessary to make some approximation. Assuming both α and λ are small and $W = C'/2$, we have

$$\frac{\partial L}{\partial h} = 0 \Rightarrow \frac{\lambda M\left(\frac{1}{p} - \frac{1}{2}\right) - \frac{\alpha C'}{h^2}}{\left[1 + \lambda\left(\frac{1}{p} - \frac{1}{2}\right)h + gn + T\right]^2} - \frac{b+Cn}{h^2} = 0.$$

Neglecting terms involving second order in α and λ

$$\begin{aligned} &\Rightarrow \quad \lambda M\left(\frac{1}{p} - \frac{1}{2}\right) - \frac{\alpha C'}{h^2} - \frac{b+Cn}{h^2} = 0 \\ &\Rightarrow \quad \lambda M\left(\frac{1}{p} - \frac{1}{2}\right) = \frac{\alpha C' + b + Cn}{h^2} \\ &\Rightarrow \quad h^2 \cong \frac{\alpha C' + b + Cn}{\lambda M\left(\frac{1}{p} - \frac{1}{2}\right)} \\ &\Rightarrow \quad h \cong \sqrt{\frac{\alpha C' + b + Cn}{\lambda M\left(\frac{1}{p} - \frac{1}{2}\right)}}. \end{aligned} \tag{2.10.15}$$

Similarly,

$$\frac{\partial L}{\partial n} = 0 \Rightarrow \left[\lambda M\left(-\frac{1}{p^2}\frac{\partial p}{\partial n}\right)h + g\right]\left[1 + \lambda\left\{\left(\frac{1}{p} - \frac{1}{2}\right)h + gn + T\right\}\right]$$

$$\frac{-\lambda\left[\left(\frac{1}{p^2}\frac{\partial p}{\partial n}\right)h + g\right]\left[\lambda W + \lambda M\left(\frac{1}{p} - \frac{1}{2}\right)h + gn + T\right] + \frac{\alpha T}{h}}{\left[1 + \lambda\left(\frac{1}{p} - \frac{1}{2}\right)h + gn' + T\right]^2} + \frac{C}{h}$$

$$\therefore L \cong \frac{\left[\lambda W + \lambda M\left(\frac{1}{p} - \frac{1}{2}\right)h + gn + T\right] + \frac{\alpha C'}{h}}{\left[1 + \lambda\left(\frac{1}{p} - \frac{1}{2}\right)h + gn + T\right] + \frac{b + Cn}{h}}$$

$$\Rightarrow \left(-\frac{\lambda M}{p^2}\frac{\partial p}{\partial n}h + g\lambda M\right) + \frac{\alpha C'}{h} + \frac{C}{h} = 0$$

Neglecting terms involving second and higher order of α and λ

$$\Rightarrow \quad -\frac{\lambda M}{p^2}\frac{\partial p}{\partial n}h^2 + g\lambda hM + \alpha C' + C = 0$$

$$\Rightarrow \quad -\frac{\lambda M}{p^2}\frac{\partial p}{\partial n}\frac{(\alpha T + b + Cn)}{\lambda M(1/p - 1/2)} + g\lambda hM + \alpha C' + C = 0.$$

On substituting the value of h^2 from (2.18)

$$\Rightarrow \quad \lambda M(\alpha T + b + Cn) = g\lambda hMp^2\frac{\partial n}{\partial p}\lambda M\left(\frac{1}{p} - \frac{1}{2}\right) + (\alpha C' + C)p^2\frac{\partial n}{\partial p}\lambda M\left(\frac{1}{p} - \frac{1}{2}\right)$$

$$\Rightarrow \quad (\alpha T + b + Cn) \cong p^2\left(\frac{1}{p} - \frac{1}{2}\right)\frac{\partial n}{\partial p}C.$$

By neglecting terms of second or higher orders of α and λ and also the terms involving $\lambda\alpha$.

$$\Rightarrow \frac{\alpha T + b}{C} + n \cong p^2\left(\frac{1}{p} - \frac{1}{2}\right)\frac{\partial n}{\partial p}. \tag{2.10.16}$$

We have

$$\begin{aligned}
\frac{\partial \pi_1}{\partial n} &= \frac{\partial}{\partial n}\frac{1}{C\lambda} = \frac{1}{\lambda}\left(-\frac{1}{C^2}\frac{\partial C}{\partial n}\right) = \frac{1}{\lambda}\left[-\frac{1}{C^2}\frac{\partial}{\partial n}\left(\frac{1}{\lambda} + \frac{h}{p} + \frac{\lambda h^2}{12} - \frac{h}{2} + gn + T\right)\right] \\
&= \frac{1}{\lambda}\left[-\frac{1}{C^2}\left(-\frac{h}{p^2}\frac{\partial p}{\partial n}\right) + g\right] = \frac{1}{\lambda}\left(\frac{h}{C^2p^2}\frac{\partial}{\partial n}(1 - \pi_1) + g\right) \\
&= \frac{1}{\lambda}\left(\frac{h}{C^2p^2}\right)\frac{\partial}{\partial n}\left[\frac{1}{\sqrt{2\pi}}\int_{-\infty}^{\sqrt{n}\delta - k} e^{-z^2/2}\,dz\right] \\
&= \frac{1}{\lambda}\left(\frac{h}{C^2p^2}\right)\frac{1}{\sqrt{2\pi}}\frac{\delta}{2\sqrt{n}}e^{-1/2(\sqrt{n}\delta - k)^2} \qquad (2.10.17) \\
\frac{\partial \pi_1}{\partial k} &= \frac{\partial}{\partial k}\frac{1}{C\lambda} = \frac{1}{\lambda}\left(-\frac{1}{C^2}\frac{\partial C}{\partial k}\right) = \frac{1}{\lambda}\left[-\frac{1}{C^2}\frac{\partial}{\partial k}\left(\frac{1}{\lambda} + \frac{h}{p} + \frac{\lambda h^2}{12} - \frac{h}{2} + gn + T\right)\right] \\
&= \frac{1}{\lambda}\left[-\frac{1}{C^2}\left(-\frac{h}{p^2}\frac{\partial p}{\partial k}\right)\right] = -\frac{1}{\lambda}\left(\frac{h}{C^2p^2}\frac{2}{\sqrt{2\pi}}e^{-k^2/2}\right) \qquad (2.10.18) \\
\frac{\partial k}{\partial n} &= \frac{\frac{1}{\lambda}\left(\frac{h}{C^2p^2}\right)\frac{1}{\sqrt{2\pi}}\frac{\delta}{2\sqrt{n}}e^{-1/2(\sqrt{n}\delta - k)^2}}{\frac{1}{\lambda}\left(\frac{h}{C^2p^2}\frac{2}{\sqrt{2\pi}}e^{-1/2(\sqrt{n}\delta - k)^2}\right)} = \frac{\delta}{2\sqrt{n}}. \qquad (2.10.19)
\end{aligned}$$

Also

$$\frac{\partial L}{\partial k} = 0 \Rightarrow M\frac{\partial}{\partial k}\left(1 - \frac{1}{C\lambda}\right) + C\frac{\partial n}{\partial k}\cdot\frac{1}{h} - \frac{W}{C^2}\frac{\partial C}{\partial k} + C'\frac{\partial}{\partial k}\left(\frac{1}{C\lambda}\right)\frac{\alpha}{h} + C'\frac{\partial}{\partial k}\frac{\alpha}{C\lambda h}.$$

Putting

$$\frac{\partial n}{\partial k} = \frac{2\sqrt{n}}{\delta}, \ \frac{\partial \alpha}{\partial k} = -\frac{2}{\sqrt{2\pi}} e^{-k^2/2}$$

$$C = \frac{1}{\lambda} + \left(\frac{h}{p} + \frac{\lambda h^2}{12} - \frac{h}{2}\right) + gn + T$$

$$C\lambda = 1 + \lambda\left(\frac{h}{p} + \frac{\lambda h^2}{12} - \frac{h}{2} + gn + T\right)$$

$$C\lambda \cong 1 \ C \cong 1/\lambda$$

$$\frac{\partial L}{\partial k} = 0 \Rightarrow \frac{C^2\sqrt{2\pi n}}{\delta C'} = e^{-k^2/2}. \tag{2.10.20}$$

Results (2.10.15), (2.10.16) and (2.10.20) have been proved by Duncan (1956).

Example 2 *Let X_2 be the time elapsed between the occurrence of an assignable cause and its detection and X_3 be the time required to inspect a sample, compute and plot its sample average and finally to find out and eliminate the assignable cause. Then,*

$$E(X_2) = \frac{h}{p} - \frac{h}{2} + \frac{\lambda h^2}{12}$$

and

$$E(X_3) = gn + T.$$

Solution: Assuming the distributions of X_2 and X_3 to be Erlangian, obtain the probability density function of $(X_2 + X_3)$.

The distribution of X_2 can approximately be taken as exponential with parameter

$$\theta_2 = \left(\frac{h}{p} - \frac{h}{2} + \frac{\lambda h^2}{12}\right)^{-1}$$

with probability density function $f_{X_2}(x_2) = \theta_2 e^{-\theta_2 x_2}$, $x_2 \geq 0$, $\theta_2 > 0$ and the distribution of X_3 to be

$$f_{X_3}(x_3) = \frac{\theta_3^r x_3^{r-1} e^{-\theta_3 x_3}}{\Gamma(r)}, \ r = 1, 2, 3, \ldots; \ x_3 \geq 0, \ \theta_3 > 0,$$

i.e., Erlangian with parameter θ_3 and r(θ_3 scale parameter and r shape parameter). Since X_2 and X_3 are independent, distribution of $X_2 + X_3 = T$ is given by

$$f_{X_2,X_2}(x_2 x_3) = f_{X_2}(x_2) f_{X_3}(x_3) = \frac{\theta_2 e^{-\theta_2 x_2} \theta_3^r x_3^{r-1} e^{-\theta_3 x_3}}{\Gamma(r)}.$$

Using the transformations $X_2 + X_3 = T$ and $V = X_2$

$$\Rightarrow x_2 = \nu, \ x_3 = t - \nu$$

$$J = \left|\frac{\partial(x_2, x_3)}{\partial(\nu, t)}\right| = \begin{vmatrix} \dfrac{\partial x_2}{\partial \nu} & \dfrac{\partial x_2}{\partial t} \\ \dfrac{\partial x_3}{\partial x_3} & \dfrac{\partial \nu}{\partial t} \end{vmatrix} = \begin{vmatrix} 1 & 0 \\ -1 & 1 \end{vmatrix} \Rightarrow |J| = 1$$

$$f_{T,V}(t,v) = \frac{\theta_2 e^{-\theta_2 \nu} \theta_3^r (t-\nu)^{r-1} e^{-\theta_3(t-\nu)}}{\Gamma(r)}, \; t \geq 0; \; 0 \leq \nu \leq t$$

$$= \frac{\theta_2 \theta_3^r (t-\nu)^{r-1} e^{-\theta_2 \nu} e^{-\theta_3(t-\nu)}}{(r-1)!}$$

$$f_T(t) = \int_0^t f_{T,V}(t,\nu)\, d\nu = \int_0^t \frac{\theta_2 \theta_3^r (t-\nu)^{r-1} e^{-\theta_2 \nu} e^{-\theta_3(t-\nu)}}{(r-1)!} d\nu$$

$$= \frac{\theta_2 \theta_3^r e^{-\theta_3 t}}{\Gamma(r)} \int_0^t e^{(\theta_3-\theta_2)\nu} (t-\nu)^{r-1}\, d\nu.$$

By repeated integration by parts, we get

$$\frac{\theta_2 \theta_3^r e^{-\theta_3 t}}{\Gamma(r)} \left[\frac{e^{t(\theta_3-\theta_2)} - 1}{(\theta_3-\theta_2)^r} - \sum_{i=1}^{r-1} \frac{t^{r-i}}{(r-1)!(\theta_3-\theta_2)^r} \right]. \tag{2.10.21}$$

Remarks

For $\theta_2 = \theta_3$ the distribution of T is Erlangian.

Example 3 *Let $Y = X_1 + T$, where X_1 is the period for which the process is in control and T is the period for which the process is not in control plus the time to detect and rectify the same. Defining $\beta(y)$ to be the time that the process is in control, show that*

$$\lim_{y\to\infty} \beta(y) = \frac{E(X_1)}{E(X_1)+E(T)} = \beta = \frac{1}{C\lambda},$$

where C is the length of complete cycle, i.e.,

$$C = \frac{1}{\lambda} + \left(\frac{h}{p} - \frac{h}{2} + \frac{\lambda h^2}{12} \right) + gn + T; \; [\textit{vide}\ (2.10.5)].$$

Solution: Note that $\beta(y)$ can be obtained as the sum of the following two probabilities, viz.,

(*i*) The probability that the assignable cause has not occurred up to y or it has occurred within the time interval y.

Let the same probability be denoted as $1 - F_{X_1}(y) = R_{X_1}(y)$, where $R_{X_1}(y)$ is the survival function and $F_{X_1}(.)$ is the cumulative distribution function for failures (assignable cause).

(*ii*) The probability that the process is in the state beyond the control in the interval $(x, x+\Delta x)$ for $x < y$ and then the state of control is restored for a time at least $(y-x)$, then, using Renewal Theory approach (*vide* Chapter 4)

$$\beta(y) = 1 - F_{X_1}(y) + \int_0^y [1 - F_{X_1}(y-x)\, dm_H(y), \tag{2.10.22}$$

where $m_H(y)$ represents the expected number of renewals in $[0, y]$ and

$$m_H(y) = \sum_{n=1}^{\infty} H_n(y),$$

where $H_n(y) = [H(y)]^n$ and $H(y)$ is the convolution distribution.

Now by the Key Renewal Theorem [*vide* page 130 and also Theorem 3.10, Ross (1970)] which states if G is a non-lattice function and

$$\int_0^y q(y)\,dy$$

is a Riemann integral, then

$$\lim_{y\to\infty}\int_0^y q(y-x)\,dm(x) = \frac{\int_0^y q(y)\,dy}{\int_0^x x dG(x)}. \tag{2.10.23}$$

Putting $G = H$ and $q(y) = 1 - F(y)$,

$$\begin{aligned}
\lim_{y\to\infty}\beta(y) &= \lim_{y\to\infty}[1 - F_{X_1}(y)] + \lim_{y\to\infty}\int_0^y [1 - F_{X_1}(y-x)]\,dm_H(y) \\
&= 0 + \lim_{y\to\infty}\int_0^y [1 - F_{X_1}(y-x)]\,dm_H(y) \\
&= \frac{\int_0^\infty [1 - F_{X_1}(y)]\,dy}{\int_0^\infty x dH(x)} \text{ by (2.10.22)} \\
&= \frac{E(X_1)}{E(Y)} = \frac{E(X_1)}{E(X_1+T)}.
\end{aligned}$$

Also,

$$\begin{aligned}
\lim_{y\to\infty}\beta(y) &= \frac{E(X_1)}{E(X_1)+E(T)} = \beta, \text{ (say)} \\
\Rightarrow \beta &= \frac{E(X)}{E(X_1)+E(X_2)+E(X_3)} = \frac{1}{\lambda C},
\end{aligned}$$

where given that C is the length of whole cycle.

2.11 Some Other Control Charts for Variables

2.11.1 Moving average charts

Suppose, $x_1, x_2, \ldots, x_p, \ldots, x_n$ be the observations at equidistant interval, then we define the moving average of order k as

$$\bar{x}_i^{(k)} = \frac{x_{i-k+1} + x_{i-k+2} + \cdots + x_i}{k}.$$

In fact, the moving average should be plotted at the middle point of the time represented (i.e., the moving average should correspond to the time in the middle interval point $[(i-k+1), i]$ or it should be $(2i - k + 1)/2$. But conventionally, the same is plotted against the line of the last (i.e., ith) observation as in the notation given. The standard error of the moving average is

$$\sigma/\sqrt{k}, \text{ where } \sigma^2 = \operatorname{var}(X_k);\ k = 1, 2, \ldots$$

The control limits are then

$$\bar{x}_i^{(k)} \pm 3.09\frac{\sigma}{\sqrt{n}}.$$

These control limits can be regarded as action limits. However, unless moving averages are separated by more than $[(i-(i-k+1)] = k-1$ sampling points, the same remain correlated, so that the control limits need not be warning limits. It has been noted by Robert (1959) while investigating ARL of moving average chart that the performance of moving average chart is much better than Shewhart's control chart; excepting that they are less quicker in detecting large changes and both of these properties become more extreme as k increases. A remedy of this deficiency is to use ordinary $\bar{X}$-chart along with a moving average chart.

Finally, moving average charts are specially suitable, where there is a time delay to produce a single item and every new observation can be added as and when it comes.

2.11.2 Moving range charts

Here we define moving ranges

$$R_i = \max(x_{i-k+1}\ldots x_i) - \min(x_{i-k+1}\ldots x_i).$$

and plot R_i over time. Action limits can be constructed accordingly. However, there is difficulty for using them as warning limits because moving ranges are in general highly correlated.

2.11.3 Geometric moving average chart

Suppose, we consider the recursive sequence (or simple Markov chain) given by

$$z_i = (1-r)z_{i-1} + rx_i$$

with $z_0 = \mu$ and $z_{-1} = 0$, μ being the target value.

$$\begin{aligned}
z_1 &= (1-r)z_0 + rx_1 = (1-r)\mu + rx_1 \\
z_2 &= (1-r)z_1 + rx_2 = (1-r)[(1-r)\mu + rx_1] + rx_2 \\
&= (1-r)^2\mu + r(1-r)x_1 + rx_2 \\
z_3 &= (1-r)^3\mu + r(1-r)^2x_1 + r(1-r)x_2 + rx_3.
\end{aligned}$$

Proceeding in this way we have

$$z_i = (1-r)^i\mu + r\sum_{k=l}^{i} x_k(1-x_k)^{i-k}.$$

$$\begin{aligned}
\text{var}(z_i) &= \sigma^2r^2\sum_{k=0}^{i-1}(1-r)^{2k} = \frac{\sigma^2r^2[1-(1-r)^{2i}]}{1-(1-r)^2} \\
&= \frac{\sigma^2r^2[1-(1-r)^{2i}]}{1-1-r^2+2r} = \frac{\sigma^2r[1-(1-r)^{2i}]}{2-r}.
\end{aligned}$$

As $i \to \infty$,

$$\text{var}(z_i) = \frac{\sigma^2r}{(2-r)},$$

which can be used as an approximation to variance except for small i. The procedure is to plot z_i against i and insert action levels as

$$\mu \pm 3.09\sigma\sqrt{r/(2-r)}.$$

For small i, the action levels should be

$$\mu \pm 3.09\sigma\sqrt{\frac{r[1-(1-r)^{2i}]}{(2-r)}}$$

Remarks

Warning limits cannot be used.

2.12 Cumulative Sum Charts (CUSUM Charts)

Suppose, we have problem of inspection of a continuous process and a batch of n items ($n \geq 1$) sampled every time at equal interval of time. The idea is to keep the process in control. For this, we require to detect the changes in the mean and the variance of the observations. This is done in all types of control charts. Shewhart's Control Chart ($\bar{X}$-chart) considers only the mean and variance of the current group of observations. On the other hand, the ordinary moving average chart as well as the Geometric moving average chart, take into consideration of the said changes of current and previous observations. But if there is a small shift in the process after some period of the beginning of the process it has been found that Shewhart's process is incapable of detecting the out of control feature of the process especially when the magnitude of the shift is relatively small. It can be shown that for small shifts the Shewhart's chart may not be effective to detect the deviation of the process from control, whereas a cumulative sum chart (CUSUM chart) is a good alternative for detecting *out of control* feature, when small shifts are important.

This is done by a CUSUM chart which directly incorporates all the information by plotting the cumulative sum of the deviations of sample values from a standard (or Target) value.

Let us take the case of a single sample observation. Suppose, x_1, x_2, ..., x_n be the observations and m be the standard or target value. Thus for preparing the CUSUM chart, we write the observations in the following scheme as follows:

Observations	Deviation of Observations from target value m	Cumulative Sum of column 2
x_1	$x_1 - m$	$x_1 - m$
x_2	$x_2 - m$	$x_1 + x_2 - 2m$
x_3	$x_3 - m$	$x_1 + x_2 + x_3 - 3m$
⋮	⋮	⋮
x_n	$x_n - m$	$\sum_{i=1}^{n} x_i - nm$

Let $E(X_i) = \mu$ then

$$E\left[\sum_{i=1}^{n}(x_i - m)\right] = n(\mu - m).$$

Therefore, the cumulative sum has a slope $(\mu - m)$ when the same is plotted against n. Hence we can interpret CUSUM as the average slope of the line

$$\sum_{i=1}^{n}(x_i - m) \text{ against } n.$$

A typical CUSUM chart is shown in Fig 12.

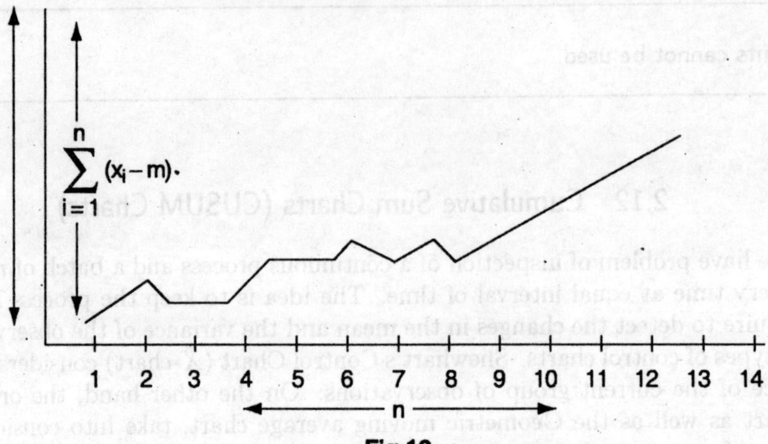

Fig 12

If $\mu = m$, the CUSUM chart will not show any slope. As soon as $\mu \neq m$ the graph will have an average slope of $(\mu - m)$.

We note that if the process remains in control with the target value m, the cumulative sum defined above is a random walk with mean zero. However, if the mean shifts to some value greater than m, then an upward or positive drift will develop in the cumulative sum. Conversely, if the mean shifts downward to some value less than m, then a downward or negative drift will develop. Therefore, if a trend develops in the plotted points either upward or downward, then the same should be considered as evidence that the process mean has a shift and a search for some assignable cause should be performed.

The theory can easily be shown by the entries in the column (2) of the above table.

Suppose, from the jth observation $(j < n)$ onwards there is a small shift in the mean in the observations. Then we like to find out how much the average slope of the CUSUM graph will be different from the average slope of the curve based on the earlier observations up to j.

Let us illustrate the concept by inserting data in the column (1) of the above table. Let the first 10 observations have been drawn from a normal population with mean $m = 10$ and standard deviation $= 1$. These observations are plotted on a Shewhart's control chart in figure 12. The central line and the 3-sigma control limits on this chart are at UCL $= 13$, CL $= 10$ and LCL $= 7$. Note that all the 10 observations plot are in control.

The last 10 observations of column (1) of the above table with data having been inserted with mean $m = 11$ and standard deviation $\sigma = 1$. Consequently, we can think of these last 10 observations having been drawn from the process which has undergone a shift in the mean by 1 time of the standard deviation. It may be seen that the last observations are not outside

Shewhart's 3-sigma control limits. So Shewhart's control chart produces no evidence of the last 10 observations. In other wards, Shewhart's control chart has failed to detect the shift.

The reason for the same is small magnitude of the shift. However, CUSUM chart in the procedure of the above tabular scheme as given below shows abrupt consistent increase from the point of shift as given in the following table.

Table 1

Sample No.	x_i	$x_i - 10$	$\sum_{i=1}^{20} x_i - 10$
1	9.380	−0.620	−0.820
2	9.696	−0.304	−0.924
3	8.584	−1.416	−2.340
4	9.423	−0.577	−2.917
5	10.524	0.574	−2.343
6	9.293	−0.707	−3.050
7	8.452	−1.548	−4.599
8	10.193	.0.193	−4.406
9	9.302	−0.698	−5.104
10	9.885	−0.115	−5.219
11	10.326	0.326	−4.893
12	11.577	1.577	−3.316
13	11.353	1.353	−1.963
14	12.056	2.056	0.093
15	11.430	1.430	1.524
16	11.656	1.656	3.180
17	11.367	1.367	4.547
18	10.016	0.016	4.563
19	10.138	0.138	4.701
20	11.686	1.686	6.387

The graph of CUSUM chart is shown in the following figure, which clearly shows that the process gets out of control while showing abrupt and consistent rise following the 10th observation onwards.

Theoretically, also we can show that if from jth sample $(j < n)$ the process begins to shift, then there is an abrupt uprise or downfall in the average slope of the CUSUM graph depending on the nature of the shift (positive or negative) from the point of shift to the final sample observations as follows.

Let

$$C_j = \sum_{i=1}^{j} (x_i - m) = (x_j - m) + \sum_{i=1}^{j-1} (x_i - m) = (x_j - m) + C_{j-1}.$$

Also,

$$C_n - C_j = \sum_{i=j+1}^{n} (x_i - m) = \sum_{i=j+1}^{n} x_i - (n-j)m$$

$$\Rightarrow \frac{\sum_{i=j+1}^{n} x_i}{n-j} = \frac{C_n - C_j}{n-j} + m = \beta(n, j) + m.$$

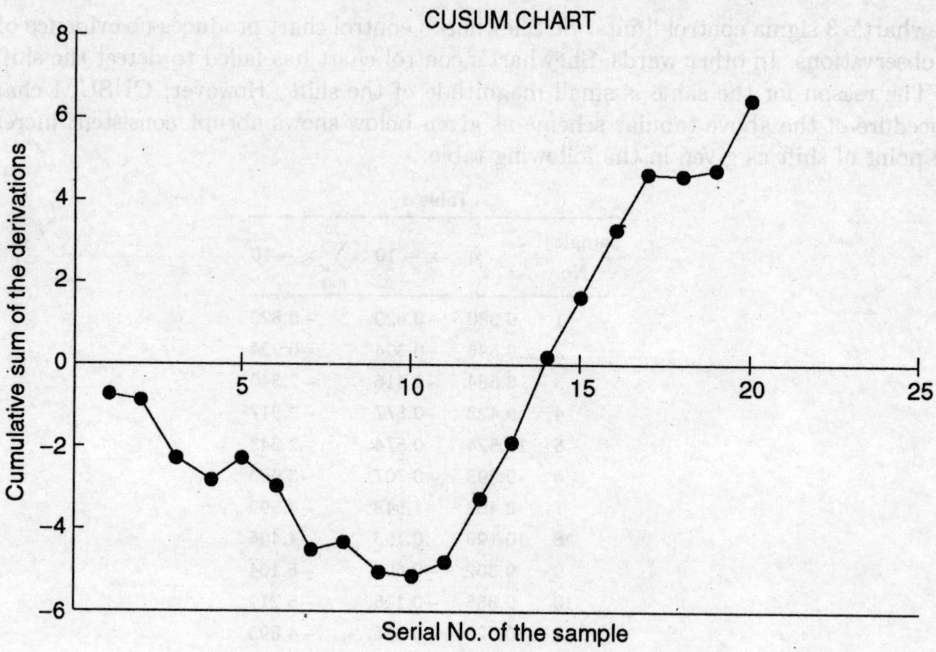

Fig 13

CUSUM Chart Crossing the Reference Value

This shows that the average value of sample number from $(j+1)$ to n is the average slope of the CUSUM graph between these two sample numbers is given by $\beta(n,j)$ plus the target value given by m. If the production process is in control, then the average value of the observations should be close to m and the slope $\beta(n,j)$ will be close to zero for $\forall\ n$ and $j(n > j)$. On the other hand if the process remains out of control because of increase or decrease of the variables of interest, then the CUSUM graph will exhibit positive or negative slope. Column (2) of the table contains the differences $(x_i - m)$ and the Column (3) gives the computation of the cumulative sum. The starting value of the CUSUM C_0 is taken as zero. Fig 12 plots the CUSUM from Column (3) of the above table.

The CUSUM chart shows an abrupt uprise or downfall if there is a sudden change in the mean in the upward or downward direction.

Note that the CUSUM plot as described above is not a control chart because it has no control limits.

2.12.1 A comparison of CUSUM chart with Shewhart's $\bar{X}$-chart

One way of comparing CUSUM chart with Shewhart's $\bar{X}$-chart is to take a few observations, *say,* the last m of the n observations and add a constant value, *say,* c to each observation in both the charts. Such a comparison has numerically been made by Wetherill (1969) and graphically illustrated. It has been shown that while X-chart shows no lack of control CUSUM chart shows it very clearly. This is given graphically in Figs 13 and 14 respectively.

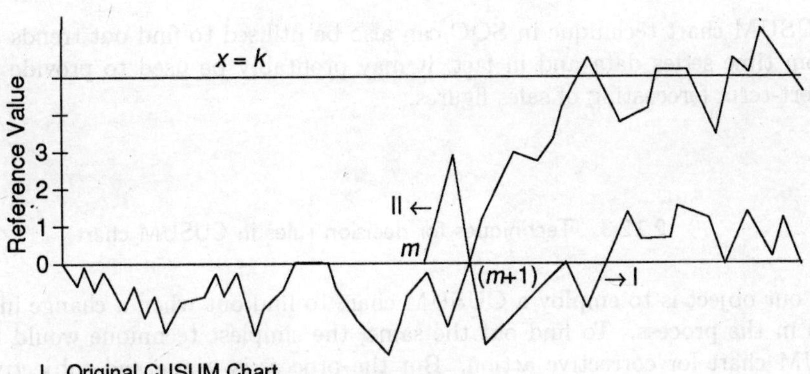

Fig 14 CUSUM Chart

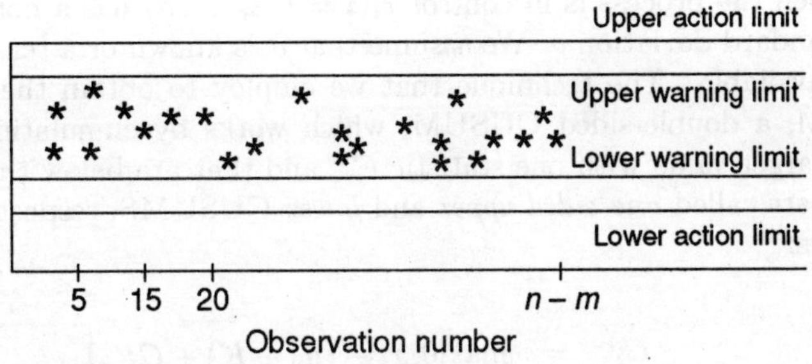

Fig 15 Shewhart's Control Chart Showing No Lack of Control

The graphs will show that small deviations from reference point will be displayed more frequently or more on the average in CUSUM chart than in Shewhart's $\bar{X}$-chart. On the other hand, large deviations from the reference values are more frequently displayed by the $\bar{X}$-chart than CUSUM chart. It also follows that small deviations are more often detected by CUSUM chart since it takes very few number of observations to see the inclination on a CUSUM chart.

2.12.2 Some applications of CUSUM chart technique

In view of the applicability of CUSUM chart and its advantages over Shewhart's $\bar{X}$ control chart we can catalogue them in the following order.

- We can detect whether there is a change in the level of mean in a continuous production process.
- While a change is detected one can also estimate the extent of change by using CUSUM chart.
- Using a CUSUM chart one can sort out continuous production process into phases, viz., effective and defective.
- CUSUM chart enables us to search when the changes in the level of mean occurred by scanning the past data.

- CUSUM chart technique in SQC can also be utilised to find out trends in sales figures from time series data and in fact, it may profitably be used to provide a technique of short-term forecasting of sales figures.

2.12.3 Techniques for decision rules in CUSUM chart

Suppose our object is to employ a CUSUM chart to find out when a change in the mean has occurred in the process. To find out the same, the simplest technique would be to examine a CUSUM chart for corrective action. But the process is vague and subjective. Therefore, more rigorous decision rule is needed to find out the above changes. Two different methods are known for the purpose, viz., (*i*) Decision interval technique, (*ii*) V-mask technique. Further, it can be shown that under certain condition both the techniques are equivalent.

(*i*) *Decision interval technique:* Let $x_1, x_2, \ldots, x_i, \ldots, x_n$ be the observations on the process. Assume that when the process is in control $x_i (i = 1, 2, \ldots, n)$ has a normal distribution with mean μ_0 and standard deviation σ. We assume that σ is known or a reasonably good estimate of the same is available. The technique that we employ to obtain the decision interval is a tabular CUSUM, a double-sided CUSUM, which works by cumulating deviations from μ_0 that are above target value with one statistic C_i^+ and that are below the target value by C_i^-. These Statistics are called *one-sided upper* and *lower* CUSUMS, respectively. C_i^+ and C_i^- are defined as follows:

$$C_i^+ = \max[0, x_i - (\mu_0 + K) + C_{i-1}^+] \qquad (2.12.1)$$

$$C_i^- = \max[0, (\mu_0 - K) - x_i + C_{i-1}^-] \qquad (2.12.2)$$

subject to $C_0^+ = C_0^- = 0$, where K is usually a quantity lying between the target value μ_0 and the out of control value of the mean given by, *say,* μ_1.

If we denote,

$$\left|\frac{\mu_1 - \mu_0}{\sigma}\right| = \delta, \text{ then } K = \left|\frac{\mu_1 - \mu_0}{2}\right| = \frac{\delta\sigma}{2}.$$

Also, denoting $K = k\sigma$, we note that deviations of observations x_i from $\mu_0 + K$ (or from $\mu_0 - K$) are cumulative only for corresponding C_i^+ or C_i^- are > 0. Otherwise, the process of making CUSUM chart begins afresh. If under this set up C_i^+ or C_i^- exceed the decision interval $I = h\sigma$ (which is conventionally taken as 5σ), then the process considered as out of control. Now for $\sigma = 1$, $K = k$ and $I = h$, we have broadly referred how to choose k and h but the proper selection of these two parameters is of considerable importance.

As the proper selection of them can only regulate the performance of the CUSUM. The procedure is illustrated in the following table constructed on the same technique as that of table 1 by generating random numbers from a normal distribution with mean = 10 and s.d. = 1 for the first 13 observations; and with a shift of mean from 10 to 11 following that while keeping the same s.d. (i.e., 1).

Table 2: $\mu_0 = 10$, $\mu_1 = 11$, $\sigma = 1$, k = 0.5 and h = 5

Period	x_i	$x_i - 10.5$	C_i^+	$9.5 - x_i$	C_i^-
1	10.130	−0.370	0	−0.630	0
2	9.829	−0.671	0	−0.329	0
3	10.643	0.143	0.143	−1.143	0
4	10.809	0.309	0.452	−1.309	0
5	9.418	−1.082	0	0.082	0.082
6	11.429	0.929	0.929	−1.929	0
7	8.484	−2.016	0	1.016	1.016
8	6.716	−3.784	0	2.704	2.704
9	10.499	−0.001	0	−0.999	1.705
10	10.661	0.161	0.161	−1.161	0.544
11	10.580	0.080	0.241	−1.080	0
12	8.668	−1.832	0	0.832	0.832
13	9.897	−0.603	0	−0.397	0.435
14	11.136	0.636	0.636	−1.636	0
15	11.605	1.105	1.741	−2.105	0
16	11.229	0.728	2.470	−1.729	0
17	13.538	3.038	5.508	−4.038	0
18	10.430	−0.070	5.438	−0.930	0
19	9.9355	0.564	6.002	−0.436	0
20	10.955	0.455	6.456	−1.455	0
21	11.672	1.726	8.182	−2.173	0
22	9.987	−0.513	7.669	−0.487	0
23	10.645	1.145	8.814	−1.157	0
24	10.081	−0.419	8.395	−0.581	0
25	12.686	2.186	10.581	−3.186	0

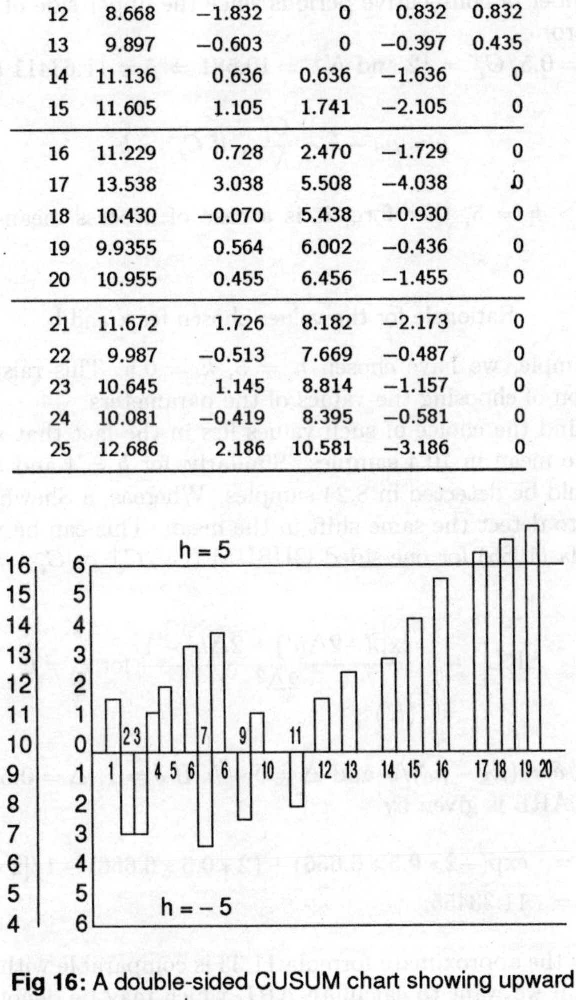

Fig 16: A double-sided CUSUM chart showing upward shift based on C_i^+ and C_i^-

Since C_i^+ has exceeded 5, the length of the decision interval $I = h$ here. Therefore, there is evidence to conclude that the process has gone out of control from the period 17 and the process was in control till the first sixteenth period. Similarly, if C_i^- exceeds the length of the decision interval, the process will be deemed to be out of control with a downfall in the CUSUM chart. Note that, the action taken following an out of control signal on a CUSUM control scheme is identical to that of any Shewhart's Control Chart.

2.13 Estimation of New Process Average when there is a Shift in the Mean detected by the CUSUM Chart

These are obtained by using the following formulas:

$$\hat{\mu} = \mu_0 + k + \frac{C_i^+}{N^+}, \text{ if } C_i^+ > h,$$

where N^+ is the number of consecutive periods since the upper side of the CUSUM C_i^+ rose above the value of zero.

Here, $\mu_0 = 10$, $k = 0.5$, $C_i^+ = 12$ and $N^+ = 10.581 \Rightarrow \hat{\mu} = 11.63411$ and

$$\hat{\mu} = \mu_0 - k - \frac{C_i^+}{N^+}, \text{ if } C_i^- > h.$$

Since C_i^- is not $> h = 5$, therefore, it is a case of process mean only going above the reference value.

Rationale for the values chosen for k and h

In the preceding example, we have chosen $h = 5$, $k = 0.5$. This raises question about the theoretical justification of choosing the values of the parameters.

The premises behind the choice of such values lies in the fact that a shift of the mean by 1σ detects shift in the mean in 10.4 samples. Similarly, for $h = 4$ and $k = 0.5$ with the same shift in the mean would be detected in 8.24 samples. Whereas, a Shewhart's chart would take about 43.96 samples to detect the same shift in the mean. This can be verified by Siegmund's approximation formula (1985) for one-sided CUSUM (i.e., C_i^+ or C_i^-) with parameters h and k given

$$\begin{aligned} \text{ARL} &= \frac{\exp(-2\Delta h') + 2\Delta h' - 1}{2\Delta^2} \text{ for } \Delta \neq 0 \\ &= (h')^2, \end{aligned}$$

where $h' = h + 1.166$, $\delta = (\mu_1 - \mu_0)/\sigma$ and $\Delta = \delta - k$. If $\delta = 1$, $\Delta = 0.5$ and $h' = 5 + 1.666 = 6.666$, then one-sided ARL is given by

$$\begin{aligned} \text{ARL} &= \exp(-2 * 0.5 * 6.666) + (2 * 0.5 * 6.666) - 1/[2 * (0.5)^2] \\ &= 11.33455. \end{aligned}$$

The value given by the approximate formula 11.33 is comparable with the actual value 10.4. For in control sample if we want to calculate ARL which may be denoted by ARL_0 we have to put $\delta = 0$ which implies $\Delta = -k$.

In this case

$$\text{ARL}_0 = \frac{\exp[-2(-0.5)(6.166) - 2(-0.5)(6.166) - 1]}{2(-0.5)^2}$$
$$= 938.2.$$

For two-sided scheme, therefore for two-sided CUSUM, the in control ARL value is given by using the formula $1/\text{ARL}_0 = 1/\text{ARL}_0^+ + 1/\text{ARL}_0^-$. Assuming $\text{ARL}_0^+ = \text{ARL}_0^-$ by symmetry, we have $1/\text{ARL}_0 = 1/938.2 + 1/938.2$ which gives $\text{ARL}_0 = 469.1$.

(*ii*) **V-mask technique:** A V-mask or a V-shaped mask is superimposed on the CUSUM chart such that the vertex of the V-mask pointing horizontally forward and is placed at a distance *d* from the last observed cumulative sum of the CUSUM chart. We denote the angle, which the oblique portion of the V-mask makes with the horizontal line by θ. If all the previously plotted points (cumulative sums) fall within the arms of the V-mask, then we *say,* that the process is in control. The parameters of the V-mask are (d, θ). Note that if the increments of a CUSUM chart make a slope, which is more than θ, then the process is said to be out of control. It may be indicative of either the process mean is higher than the reference value or lower than the same. The concepts are diagrammatically illustrated below in Figs 17, 18 and 19.

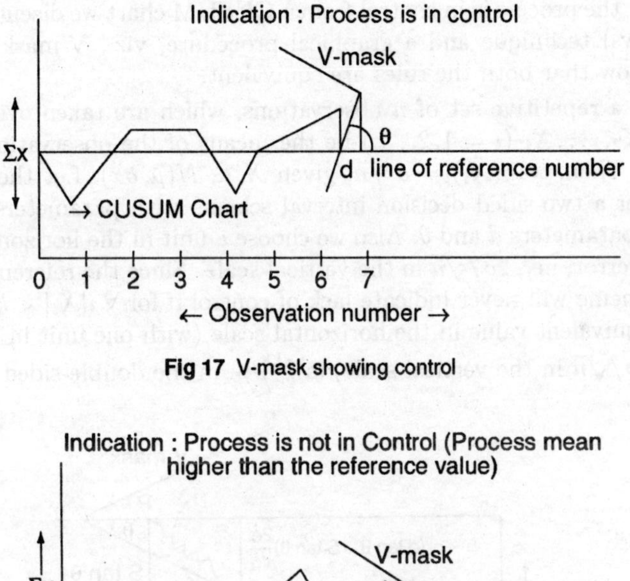

Fig 17 V-mask showing control

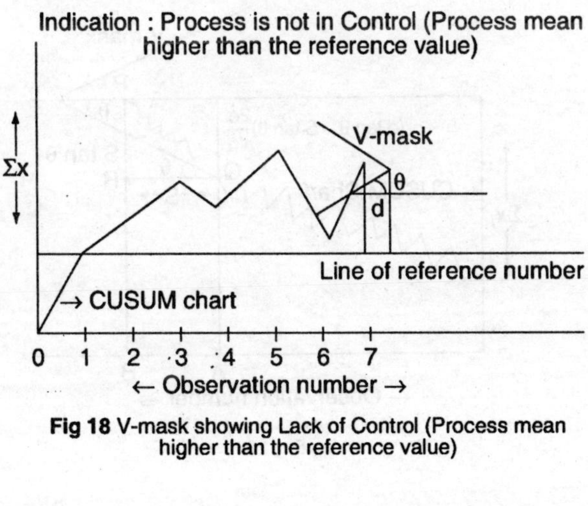

Fig 18 V-mask showing Lack of Control (Process mean higher than the reference value)

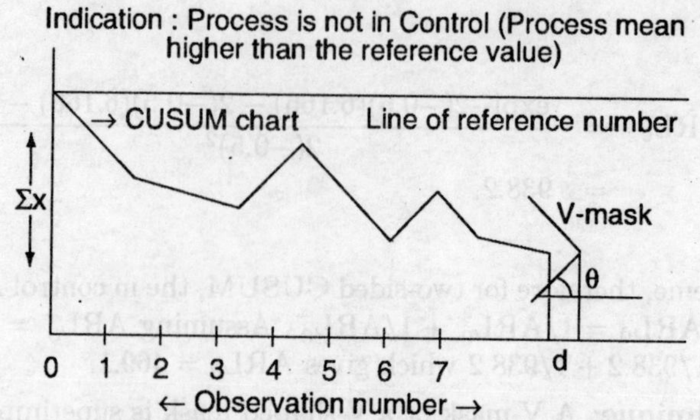

Fig 19 V-mask showing Lack of Control (Process mean lower than the reference value)

It may be seen that the maximum inclination of the CUSUM chart $< \theta$ in Fig 17, whereas in Figs 18 and 19 the maximum inclination exceeds θ. The parameters of the V-mask (d, θ) controls the property of the V-mask. The concept of V-mask was introduced by Barnard (1959).

2.14 Equivalence of Two Decision Rules

To find out whether the process is in control from a CUSUM chart we discussed two procedures, viz., decision interval technique and a graphical procedure, viz., V-mask technique. In this section, we shall show that both the rules are equivalent.

Let us consider a repetitive set of n observations, which are taken over equal intervals of time; and let $\bar{X}_1, \bar{X}_2, \ldots, \bar{X}_i$ $(i = 1, 2, \ldots)$ be the means of the observations each time based on sample sizes n. Then, var $(\bar{X}_i) = \sigma^2/n$, given $X \simeq N(\mu, \sigma^2)$. Let the reference value be $m = 0$. We consider a two-sided decision interval scheme with parameters h and k and a V-mask scheme with parameters d and θ. Also we choose a unit in the horizontal scale, equal to 2 times the standard error, i.e., $2\sigma/\sqrt{n}$ in the vertical scale. Since the reference value $m = 0$, the decision interval scheme will never indicate lack of control if for $\forall\, i |\bar{X}_i| < k$. Since $\bar{X}_i$ is in the vertical scale, its equivalent value in the horizontal scale (with one unit in the horizontal scale corresponding to $2\sigma/\sqrt{n}$ in the vertical scale) is $\frac{|\bar{X}_i|}{2\sigma/\sqrt{n}}$ in the double-sided decision scheme.

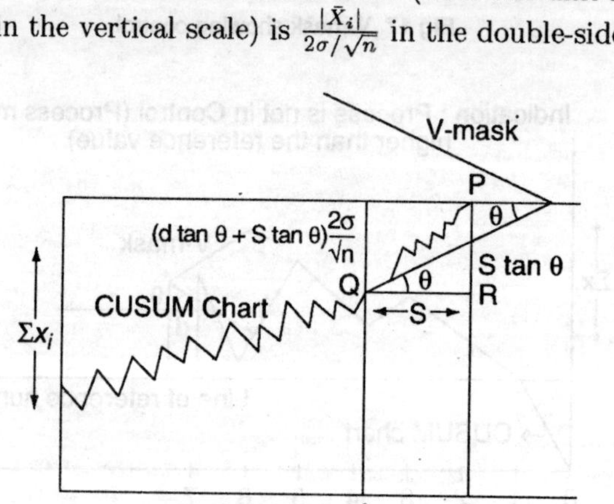

Fig 20

Therefore, $\frac{|\bar{X}_i|}{2\sigma/\sqrt{n}}$ representing the increments to the CUSUM chart should not cause a slope (which is measured from a line parallel to horizon) more extreme than that of the V-mask. More precisely,

$$\frac{|\bar{X}_i|}{2\sigma/\sqrt{n}} < \tan\theta.$$

Therefore, if we put $k = (2\sigma/\sqrt{n})\tan\theta$, the restrictions, in the decision interval scheme and the V-mask will be equivalent.

The V-mask will indicate a need for corrective action at a point P (*vide* Fig 20) in case there is a point Q crossing one of the arms of the V-mask. This implies that the summation of $\bar{X}_i$ from A to B exceeds PR. In other wards,

$$\sum_A^B \bar{X}_i > \mathrm{PR} = (d\tan\theta + s\tan\theta)\frac{2\sigma}{\sqrt{n}} \tag{2.14.1}$$

$$\Rightarrow \quad \sum_A^B \left[\bar{X}_i - \frac{2\sigma}{\sqrt{n}}\tan\theta\right] > (d\tan\theta)\frac{2\sigma}{\sqrt{n}} \tag{2.14.2}$$

since the measure of AB is s.

If we put

$$\frac{2\sigma}{\sqrt{n}}\tan\theta = k \text{ and } \frac{2\sigma}{\sqrt{n}} d\tan\theta = h. \tag{2.14.3}$$

Since the equation (2.13.2) defines a decision interval scheme, therefore, by applying V-mask technique we arrive at the same. This shows the equivalence of two procedures.

2.14.1 V-mask procedure

Note that for the equivalence of decision interval and V-mask techniques, the parameters of the V-mask viz., d and θ are connected with the decision interval scheme, viz., h and k by the following

$$h = dk \text{ and } \theta = \tan^{-1}\frac{k}{A}, \tag{2.14.4}$$

where A is the horizontal distance on the V-mask plot between successive points in terms of unit distance on the vertical scale. In that case,we may write

$$k = A\tan\theta \text{ and } h = Ad\tan\theta. \tag{2.14.5}$$

Now A is comparable with $\frac{2\sigma}{\sqrt{n}}$ in (2.13.3). Putting $\sigma = 1$ and $n = 1$, we get $A = 2$. Choosing $k = 1/2$ and $h = 5$ as in the decision interval technique illustrating problem in table 2, we get, $\theta = \tan^{-1}(1/4) = 14.036°$ and $d = h/k = 10$ which gives the parameters of the V-mask from the known parameters of the equivalence of decision interval scheme. Johnson (1961) has suggested a method for designing the V-mask for selecting the parameters d and θ. This is given by

$$\theta = \tan^{-1}\left(\frac{\delta}{2A}\right) \text{ and } d = \left(\frac{2}{\delta^2}\right)\log_e\left(\frac{1-\beta}{\alpha}\right), \tag{2.14.6}$$

where 2α is the greatest permissible probability of a signal when the process mean is on target (i.e., false alarm) and β is the probability of not detecting a shift of size δ.

For small β

$$d = -\log \alpha / \ddot{o}. \tag{2.14.7}$$

Many computer programs are based on Johnson's design method. Usual default settings of α, β and $\ddot{o}$ are chosen as 0.05, 0.05 and 1 respectively.

2.14.2 A numerical illustration of a Double-sided CUSUM showing lack of control for downward shift

Table 3: A double-sided CUSUM chart showing lack of control for downward shift

x_i	$[x_i - 24]$	CUSUM	Decision
23.43796	−0.5620355	−0.5620	
25.19394	1.1939403	0.6319	
24.61993	0.6199271	1.2518	
29.29302	5.2930152	6.5448	
24.79560	0.7956041	7.3404	Decision boundary not crossed
21.78811	−2.2118920	5.1285	
23.77563	−0.2243733	4.9042	
23.75065	−0.2493526	5.1536	
21.33277	−2.6672305	2.4864.	
20.16319	−3.8368058	−1.3504	
24.79713	0.7971347	−0.5533	
21.62774	−2.3722591	−2.9255	
23.01007	−0.9899330	−3.9154	
21.52657	−2.4734327	−6.3888	
24.65215	0.6521504	−5.9367	
24.57684	0.5768394	−5.1599	
24.10438	0.1043764	−5.0555	
21.10610	−2.8938972	−7.9493	
22.87729	−1.1227146	−9.0720	
18.73381	−5.2661884	−14.6381	Decision boundary crossed
23.25929	−0.7407144	−15.0788	
26.63506	2.6350599	−12.4438	
20.00921	−3.9907852	−16.4346	
22.85298	−1.1470232	−17.5816	

Here $\sigma = 2$, Subgroup size n, $n = 1$, $I =$ Decision boundary, $h_\sigma = 5 \times 2 = 10$ and $K = 1$, $\mu_0 = 25$, $\mu_1 = 23$. Here for lower-sided value K is replaced by $-K = -1$ and I is replaced by $-I = -10$. The two-sided CUSUM test reveals that the process has a lack of control in the downward direction. Further, the Shewhart's control chart give

$$\text{UCL} = \mu_0 + 3\sigma/\sqrt{n} = 25 + 3.2 \simeq 31.$$

And

$$\text{LCL} = \mu_0 - 3\sigma/\sqrt{n} = 25 - 3.2 = 19.$$

It shows that Shewhart's control chart shows no lack of control but CUSUM chart does.

2.15 Tolerance Range

The concept of tolerance is distinct from that of control limits or confidence intervals. The confidence interval is said to include a population parameter with some probability statement, whereas a tolerance range is said to include a certain percentage or more of the population with some probability statements.

Example 4 *100β% confidence interval of a population parameter m in a normal population may be set as $\bar{X} \pm CS$. This means that in β proportion of samples, the confidence interval will include the population parameter with certainty. While 100γ% of the tolerance range with confidence coefficient β may be set as $\bar{X} \pm KS$. This means that in β proportion of samples, the tolerance range will include 100γ% or more of the population. C and K are both independent of population parameters. C is a function of β and n (the sample size), whereas K is a function of β, γ and n.*

Solution: When the population parameters of a normal population m and σ^2 known we can set 100γ% tolerance range as $m \pm K\sigma$ such that

$$\int_{m-K\sigma}^{m+K\sigma} \frac{1}{\sigma\sqrt{2\pi}} e^{-(x-m)^2/2\sigma^2}\, dx = \gamma.$$

But in almost all situations m and σ^2 need not be known. In control charts, we have replaced m and σ^2 by their unbiased estimates. It is evident that as $\bar{X}$ and S vary from sample to sample, the proportion of population included in the range $(\bar{X} \pm KS)$ will also be a random variable varying from sample to sample. More precisely,

$$\int_{\bar{X}-KS}^{\bar{X}+KS} \frac{1}{\sigma\sqrt{2\pi}} e^{-(x-m)^2/2\sigma^2}\, dx$$

ceases to take a fixed value equal to γ but is a random variable. In this situation it is not possible to get an interval covering a fixed proportion of population. What at best can be made is to find out an interval $\bar{X} \pm KS$ such that the probability is β that 100γ% or more of the population is included in the range and the same way be called as 100γ% *tolerance range* (or *confidence range*) with confidence coefficient β.

Let

$$\int_{\bar{X}-KS}^{\bar{X}+KS} \frac{1}{\sigma\sqrt{2\pi}} e^{-(x-m)^2/2\sigma^2}\, dx = P.$$

Then we are to obtain K so that

$$P(P \geq \gamma) = \beta \Rightarrow \int_{\gamma}^{1} f(P)\, dP = \beta,$$

where $f(P)$ is the probability density function of P.

The distribution of P is quite complicated and not exactly known. Wald and Wolfwitz (1946) have given an approximate solution. According to them the value of K is given by

$$K = r\sqrt{\frac{n-1}{\chi^2_\beta}}, \tag{2.15.1}$$

where χ^2_β is $100\beta\%$ critical value of χ^2 with $(n-1)$ degrees of freedom and is given by

$$\frac{1}{\sqrt{2\pi}}\int_{(1/\sqrt{n})-r}^{(1/\sqrt{n})+r} e^{-\xi^2/2}\,d\xi=\gamma. \tag{2.15.2}$$

The tolerance limits are simple and is a variable by using χ^2-table and Normal probability table. Wald and Wolfwitz have also examined the order of accuracy of the approximate tolerance as given above and found the results are fairly accurate.

A next possibly best approach is to construct average tolerance range. Given $\bar{X} \pm KS$ as the average tolerance range such that the same range will contain $100\alpha\%$ of the populations on the average

$$\text{or, } E(P)=\alpha=\frac{\Gamma\left(\frac{n}{2}\right)}{(\sqrt{n-1})\,\Gamma\left(\frac{n-1}{2}\right)\Gamma\left(\frac{1}{2}\right)}\int_{-t}^{t}\frac{d\xi}{1+\left(\frac{\xi^2}{n-1}\right)^{n/2}}, \tag{2.15.3}$$

where

$$t=K\sqrt{\frac{n}{n+1}}.$$

Also

$$E(P)=\int_{-\infty}^{\infty}\int_{0}^{\infty} f(\bar{x},s)P\,d\bar{x}ds=\alpha,$$

where $f(\bar{x},s)$ is the joint probability density function of $\bar{X}$ and S. $1-\alpha$ is the value of the two-sided level of significance in the t-table with $(n-1)$ distribution function corresponding to $t=K\sqrt{n}/(n+1)$, where n is the sample size; Wilks, S. S. (1941).

Thus if we are to calculate the tolerance range such that the average proportion of population included in the range is $100\alpha\%$, we set the same as

$$\bar{X}\pm KS, \text{ where } K=\sqrt{\frac{n+1}{n}}t_{1-\alpha};$$

$t_{1-\alpha}$ being the $100(1-\alpha)\%$ value of t with $(n-1)$ degrees of freedom.

The above discussion is confined to normal population only; and asymptotic solution when the functional form of the population is known; but the parameters involved are unknown and have been obtained by Wald (1942). However, the same are applicable only in large samples.

The problem discussed above can be simplified to a great extent if one of the two parameters m and σ^2 in a Normal population is known.

2.16 Tolerance Range based on Ordered Statistics

2.16.1 Tolerance range based on two extreme observations $x_{(1)}$ and $x_{(n)}$

The joint distribution of $x_{(1)}$ and $x_{(n)}$ is given by

$$\frac{n!}{1!(n-2)!1!}\left[\int_{x_{(1)}}^{x_{(n)}} f(x)\,dx\right]^{n-2} f(x_{(1)})f(x_{(n)})\,dx_{(n)}\,dx_{(n)}.$$

Putting

$$\int_{-\infty}^{x_{(l)}} f(x)\,dx = u \text{ and } \int_{x_{(1)}}^{x_{(n)}} f(x)\,dx = v,$$

we have

$$|J| = \frac{\partial(x_{(1)}x_{(n)})}{\partial(u,v)} \equiv 1 \Big/ \frac{\partial(u,v)}{\partial(x_{(1)},x_{(n)})} = \frac{1}{f(x_{(i)})f(x_{(n)})}$$

Hence the distribution of u and v is given by

$$n(n-1)v^{n-2}\,du\,dv;\ 0 \le u \le 1-v;\ 0 \le v \le 1.$$

The region of non-zero probability density is the triangle bounded by the lines $u = 0$, $v = 0$ and $u + v = 1$.

Integrating with respect to u from 0 to $(1-v)$ we get the probability distribution of v as

$$f(v) = \begin{cases} n(n-1)v^{n-2}\displaystyle\int_0^{1-v} du \\ n(n-1)v^{n-2}\,(1-v). \end{cases} \tag{2.16.1}$$

We have

$$n(n-1)\int_\gamma^1 v^{n-2}\,(1-v)\,dv = \beta$$

is the $100\gamma\%$ tolerance range with confidence coefficient β.

$$\Rightarrow \quad n(n-1)\left[\frac{v^{n-1}}{n-1} - \frac{v^n}{n}\right]_\gamma^1 = \beta$$

$$\Rightarrow \quad n(1-\gamma^{n-1}) - (n-1)(1-\gamma^n) = \beta$$

$$\text{or,} \quad n\gamma^{n-1} - (n-1)\gamma^n = 1 = \beta.$$

Given β and γ, one can solve n by successive iteration, *say,* by Newton-Raphson method.

Example 5 *Given* $\gamma = 0.95$ *and* $\beta = 0.99$, *obtain* n *so that* $(x_{(1)}x_{(n)})$ *is 95% tolerance interval with confidence coefficient 99%.*

Solution: $n = 130$.

2.16.2 Distribution of tolerance range based on $x_{(r)}, x_{(n-r+1)}$, where $x_{(1)} < x_{(2)} < \ldots x_{(r)} < \cdots < x_{(n-r+1)} < \cdots < x_{(n)}$

Let

$$P = \int_{x_{(r)}}^{x_{(n-r+1)}} f(x)\,dx,\ p_r = \int_{-\infty}^{x_{(r)}} f(x)\,dx \text{ and } p_{n-r+1} = \int_{-\infty}^{x_{(n-r+1)}} f(x)\,dx,$$

then $f(x_{(r)})\,dx_{(r)} = dp_r$ and $f(x_{(n-r+1)}\,dx_{(n-r+1)}) = dp_n$.

Joint distribution of $x_{(r)}$ and $x_{(n-r+1)}$ is

$$\frac{n!}{[(r-1)!]^2(n-2r)!}\left[\int_{-\infty}^{x_{(r)}} f(x)\,dx\right]^{r-1}\left[\int_{x_{(r)}}^{x_{(n-r+1)}} f(x)\,dx\right]^{n-2r}\left[\int_{x_{(n-r+1)}}^{\infty} f(x)\,dx\right]^{r-1}$$

$$\times f(x_{(r)})f(x_{(n-r+1)})\,dx_{(r)}\,dx_{(n-r+1)}$$

$$= \frac{n!}{[(r-1)!]^2(n-2r)!}p_r^{r-1}(p_{n-r+1}-p_r)^{n-2r}(1-p_{n-r+1})^{r-1}\,dp_r\,dp_{n-r+1}.$$

Let $Q = p_r$ and $P = p_{n-r+1} - P_r$. Hence the joint distribution of P and Q is

$$dF = \frac{n!}{[(r-1)!]^2(n-2r)!} Q^{r-1} p^{n-2r} (1-P-Q)^{r-1}\, dP dQ.$$

The distribution of P is given by $f(P)\, dP$

$$\begin{aligned} f(P)\, dP &= \frac{n!}{[(r-1)!]^2(n-2r)!} \int_0^{1-p} Q^{r-1} p^{n-2r} (1-P-Q)^{r-1}\, dQ dP \\ &= \frac{n! P^{n-2r}}{[(r-1)!]^2(n-2r)!} \int_0^{1-P} \left(1 - \frac{Q}{1-P}\right)^{r-1} \frac{dQ}{1-P} (1-P)^{2r-1}\, dP. \end{aligned}$$

Putting

$$\begin{aligned} R = \frac{Q}{(1-P)} &= \frac{n!(1-P)^{2r-1} P^{n-2r}\, dP}{[(r-1)!]^2(n-2r)!} \int_0^1 R^{r-1} (1-R)^{r-1}\, dR \\ &= \frac{n!(1-P)^{2r-1} P^{n-2r}\, dP}{[(r-1)!]^2(n-2r)!} \frac{(r-1)!(r-1)!}{(2r-1)!} \\ &= \frac{\Gamma(n+1)}{\Gamma(2r)\Gamma(n-2r+1)} P^{n-2r} (1-P)^{2r-1}\, dP \\ &= \frac{1}{B(n-2r+1, 2r)} P^{n-2r} (1-P)^{2r-1}\, dP. \end{aligned} \tag{2.16.2}$$

Example 6 *Using the result*

$$\frac{1}{B(n-2r+1, 2r)} \int_\gamma^1 P^{n-2r} (1-P)^{2r-1}\, dP = \beta,$$

which gives $100\gamma\%$ *tolerance range with confidence coefficient* β *based on* $(x_{(r)}, x_{(n-r+1)})$. *Obtain* n *(given* r*) using Pearson's incomplete Beta function table. Alternatively,*

$$n = \frac{1}{4}\chi^2_{1-\beta,2r} \cdot \frac{1+r}{1-r} + \frac{1}{2}(2\gamma - 1),$$

where $\chi^2_{1-\alpha,2r}$ *is* $100(1-\beta)\%$ *value of* χ^2 *with* $2r$ *degrees of freedom.*

Example 7 *Obtain the tolerance range based on* $(x_{(r)}, \infty)$.

Solution: Let

$$P = \int_{x_{(r)}}^\infty f(x)\, dx = 1 - P_r.$$

Distribution of p_r is given by

$$\frac{n!}{(r-1)!1!(n-r)!} p_r^{r-1} (1-p_r)^{n-r}\, dp_r,$$

where $\dfrac{dp_r}{dx_{(r)}} = f(x_{(r)})$.

Hence the distribution of P is given by

$$\begin{aligned} f(P)\, dP &= \frac{\Gamma(n+1)}{\Gamma(r)\Gamma(n-r+1)} P^{n-r} (1-P)^{r-1}\, dP \\ &= \frac{1}{B(n-r+1, r)} P^{n-r} (1-P)^{r-1}\, dP. \end{aligned} \tag{2.16.3}$$

Remarks

Note that (2.14.3) is obtainable from (2.14.2) by replacing 2r by r.

$100\gamma\%$ tolerance range with confidence coefficient β is given by

$$\frac{1}{B(n-r+1,r)}\int_\gamma^1 P^{n-r}\,(1-P)^{r-1}\,dP=\beta.$$

Again given β and γ, n is obtainable by using Pearson's incomplete Beta function table.

Example 8 *Obtain the tolerance interval based on* $(-\infty, x_{(n-r+1)})$.

Solution: Hence we set

$$P=\int_{-\infty}^{x_{(n-r+1)}} f(x)\,dx=p_{n-r+1}.$$

Distribution of p_{n-r+1} is given by

$$F=\frac{n!}{(r-1)!1!(n-r)!}\,(p_{n-r+1})^{n-r}\,(1-p_{n-r+1})^{r-1}\,dp_{n-r+1}.$$

Hence the distribution of P is given by

$$\begin{aligned} f(P)\,dP &= \frac{\Gamma(n+1)}{\Gamma(r)\Gamma(n-r+1)}\,P^{n-r}\,(1-P)^{r-1}\,dP \\ &= \frac{1}{B(n-r+1,r)}\,P^{n-r}\,(1-P)^{r-1}\,dP. \end{aligned} \tag{2.16.4}$$

$100\gamma\%$ tolerance with confidence coefficient β is given by

$$\frac{1}{B(n-r+1,r)}\int_\gamma^1 P^{n-r}\,(1-P)^{r-1}\,dP=\beta. \tag{2.16.5}$$

Using Incomplete Beta function table, we have

$$I_\gamma(n-r+1,r)=1-\beta. \tag{2.16.6}$$

Scheffe and Tukey (1945) have given an approximate solution for n as

$$n=\frac{1}{4}\chi^2_{1-\beta,2r}\,\frac{1+r}{1-r}+\frac{1}{2}\,(2\gamma-1), \tag{2.16.7}$$

where $\chi^2_{1-\beta,2r}$ is $100(1-\beta)\%$ value of χ^2 with $2r$ degrees of freedom.

2.17 Normal Tolerance Range

2.17.1 $100\gamma\%$ tolerance range with confidence coefficient β when μ (mean) is known but σ (standard deviation) is unknown

Case I *Here we estimate* σ^2 *by*

$$s^2=\sum_{i=1}^{n}\frac{(x_i-\bar{x})^2}{n-1},$$

where $\bar{x}$ *is the sample mean* (s^2 *being the unbiased estimate of* σ^2).

We set the tolerance interval as $(\mu - ks,\ \mu + ks)$ for some k (to be obtained) such that $(\mu - ks,\ \mu + ks)$ will contain at least $100\gamma\%$ population with probability β. The levels of γ and β are given. The idea is to consider fixed interval of the form $(\mu - k'\sigma,\ \sigma + k'\sigma)$ so that $(\mu - ks, \mu + ks)$ is larger than $(\mu - k'\sigma, \mu + k'\sigma)$ in $100\beta\%$ of the cases. Also set k' in such a way that $(\mu - k'\sigma, \mu + k'\sigma)$ will exactly contain $100\gamma\%$ of the population with probability unity; or $(\mu - ks, \mu + ks)$ will be $100\gamma\%$ tolerance interval with confidence coefficient β. Then the required tolerance range is given by $\mu \pm ks$, where k is so chosen that

$$\mu \pm k\sqrt{\frac{c}{n-1}}\sigma \tag{2.17.1}$$

will include $100\gamma\%$ of the population and c being determined from the following.

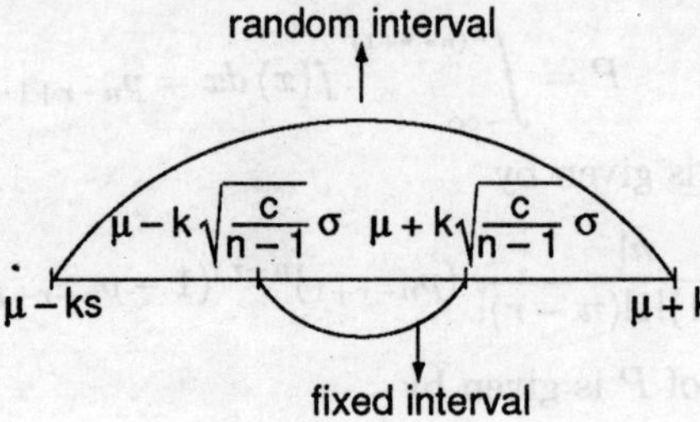

Fig 21

$$P\left(\frac{(n-1)s^2}{\sigma^2} \geq c\right) = \beta$$

$$\Rightarrow \quad P\left(s \geq \sqrt{\frac{c}{n-1}}\sigma\right) = \beta$$

$$\Rightarrow \quad P(\chi^2_{n-1} \geq c) = \beta \tag{2.17.2}$$

$$\Rightarrow \quad P\left(\mu + ks \geq \mu + k\sqrt{\frac{c}{n-1}}\sigma = \mu + k'\sigma \cup \mu - ks \leq \mu - k\sqrt{\frac{c}{n-1}}\sigma = \mu - k'\sigma\right) = \beta$$

$$P((\mu - ks, \mu + ks) \geq (\mu - k'\sigma, \mu + k'\sigma)) = \beta,$$

where $k' = k\sqrt{\dfrac{c}{n-1}}$.

Also

$$P\left(\mu - k\sqrt{\frac{c}{n-1}}\sigma, \mu + k\sqrt{\frac{c}{n-1}}\sigma\right) = \gamma.$$

Case II *If σ^2 is estimated as*

$$s_1^2 = \sum_{i=1}^{n} \frac{(x_i - \mu)^2}{n}$$

(s_1^2 is also an unbiased estimate of σ^2). Tolerance range is $\mu \pm ks_1$, where k is so chosen that

$$\mu \pm k\sqrt{\frac{c}{n}}\sigma \tag{2.17.3}$$

will include 100γ% of the population and c being obtained from the following.

$$P\left(\frac{ns_1^2}{\sigma^2} \geq c\right) = \beta \Rightarrow P(\chi_n^2 \geq c) = \beta, \tag{2.17.4}$$

where c is determined from χ^2-table (100β% point) with n degrees of freedom and k is obtainable from Normal distribution table as before.

Case III *If σ^2 is estimated as*

$$s_2^2 = \sum_{i=1}^{n} \frac{(x_i - \bar{x})^2}{n}$$

(s_2^2 is not an unbiased estimate of σ^2), then tolerance range is given by $\mu \pm ks_2$, where k is so chosen that

$$\mu \pm k\sqrt{\frac{c}{n}}\sigma \tag{2.17.5}$$

will include 100γ% of the population and c being obtained from the following.

$$P\left(\frac{ns_2^2}{\sigma^2} \geq c\right) = \beta, \text{ or, } P(\chi_{n-1}^2 \geq c) = \beta, \tag{2.17.6}$$

where c is obtained from χ^2-table with $(n-1)$ degrees of freedom and k is obtainable from Normal distribution table.

2.17.2 100γ% tolerance range with confidence coefficient β when σ is known but μ is unknown

The Tolerance Range (T.R.) is given by $\bar{x} \pm k\sigma$, where μ is estimated by $\bar{x}$ and k is so chosen that

$$\left(\mu + \sqrt{\frac{d}{n}}\sigma\right) \pm k\sigma \tag{2.17.7}$$

will include 100γ% of the population. Also $\bar{x} + k\sigma$ will be larger than

$$\left(\mu + \sqrt{\frac{d}{n}}\sigma\right) \pm k\sigma$$

in 100β% of the cases making $(\bar{x} - k\sigma, \bar{x} + k\sigma)$ a tolerance interval with confidence coefficient β subject to

$$P\left(\frac{|\bar{x} - \mu|}{\sigma/\sqrt{n}} \geq d\right) = \beta \tag{2.17.8}$$

$$\Rightarrow \text{ either } P(\tau \geq d) = \beta, \text{ where } \tau = \frac{|\bar{x} - \mu|}{\sigma/\sqrt{n}} \sim N(0,1)$$

$$\Rightarrow P(\bar{x} \geq \mu + d\sigma/\sqrt{n}) = \beta$$

$$\Rightarrow P(\bar{x} + k\sigma \geq \mu + k\sigma + d\sigma/\sqrt{n}) = \beta \tag{2.17.9}$$

$$P\left(\frac{\bar{x} - \mu}{\sigma/\sqrt{n}} \leq -d\right) = \beta \tag{2.17.10}$$

$$P(\bar{x} \leq \mu - d\sigma/\sqrt{n}) = \beta$$
$$\Rightarrow \quad P(\bar{x} - k\sigma \leq \mu - k\sigma - d\sigma/\sqrt{n}) = \beta. \tag{2.17.11}$$

Combining (2.17.9) and (2.17.11)

$$P(\bar{x} - k\sigma \leq \mu - \sigma(k + d/\sqrt{n}) \cup \mu + \sigma(k + d/\sqrt{n}) \leq \bar{x} + k\sigma) = \beta$$
$$P(\bar{x} - k\sigma \leq \mu - k'\sigma \cup \mu + k'\sigma \leq \bar{x} + k\sigma) = \beta, \tag{2.17.12}$$

when $k' = k + d/\sqrt{n}$.

Suppose,

$$P(\mu - \sigma(k + d/\sqrt{n}),\ \mu + \sigma(k + d/\sqrt{n}) = \gamma,$$

then $P[(\bar{x} - k\sigma, \bar{x} + k\sigma) \geq \gamma] = \beta$, i.e., $(\bar{x} - k\sigma, \bar{x} + k\sigma)$ is $100\gamma\%$ tolerance range with confidence coefficient β. Both d and k are obtainable from Normal distribution table.

2.17.3 100γ% tolerance range when confidence coefficient β with both μ and σ are unknown

Here μ is estimated by $\bar{x}$ and σ^2 by s^2, where

$$s^2 = \sum_{i=1}^{n} \frac{(x_i - \bar{x})^2}{n-1}.$$

Tolerance range is given by $\bar{x} \pm ks$, where

$$k = r\sqrt{\frac{n-1}{\chi^2_{n-1}(\beta)}}, \tag{2.17.13}$$

where r is chosen that

$$\frac{1}{\sqrt{2\pi}} \int_{c-r}^{c+r} e^{-\tau^2/2}\, d\tau = r,$$

where $c = 1/\sqrt{n}$ and $\chi^2_{n-1}(\beta)$ is $100\beta\%$-value of χ^2 with $(n-1)$ distribution function.

2.17.4 Tolerance range when the average population included is 100α%; μ and σ both unknown

Here

$$\text{T.R.} = \bar{x} \pm ks, \text{ where } k = \sqrt{\frac{n-1}{n}}\, t_{n-1,(1-\alpha)}, \tag{2.17.14}$$

and $t_{n-1,(1-\alpha)}$ is the two-sided $100(1-\alpha)\%$ critical value of Student's t distribution with $(n-1)$ degrees of freedom.

Example 9 *Obtain the average tolerance interval when both mean and variance are not known.*

Solution: Let X be an independent observation other than $X_1, X_2, \ldots, X_N$.

$$\bar{X} = \Sigma X_i/n,$$

so that $E_\theta(X - \bar{X}) = 0$; $\text{var}_\theta(X - \bar{X}) = \sigma^2 + \sigma^2/n$; $\text{cov}_\theta(X - \bar{X}) = 0$

$$t = \frac{(X - \bar{X}) - 0}{s\sqrt{1 + 1/n}},$$

where s^2 is unbiased for σ^2

$$P\left(-t_{\alpha/2,n-1} \le \frac{X-\bar{X}}{s\sqrt{1+1/n}} \le t_{\alpha/2,n-1}\right) = 1-\alpha$$

$$\Rightarrow\ P\left[\bar{X} - t_{\alpha/2,n-1}\ s\sqrt{1+\frac{1}{n}} \le \bar{X} \le \bar{X} + t_{\alpha/2,n-1}\ s\sqrt{1+\frac{1}{n}}\right] = 1-\alpha.$$

Let ξ denote a random variable $\ni$

$$\begin{aligned}\xi &= 1 \text{ with probability } (1-\alpha)\text{, i.e., when the interval covers } X\\ &= 0\text{, when the interval does not cover } X.\end{aligned}$$

$E(\xi) = (1-\alpha)$, i.e., on the average in only $(1-\alpha)\%$ of the cases in the interval covers the population.

Chapter 3

Reliability

3.1 Introduction

Almost everyone of us is acquainted with the term *Reliability* in our daily life. When we assign the attribute *reliable* to a component or a system (a system may be concerned as the collection of several components), we precisely mean to say that the same will render service for a good or at least reasonable period of time. When we come to quantitative terms to estimate or measure the reliability of a component or a system, we indeed come to the concept of chance or probability that a component or a system survives up to a given length of time.

Thus we define Reliability as the probability of a system (or a component) surviving at least up to certain length of time. Denoting T, the random variable corresponding to survival or longevity of a component or system, we have, $R(t)$, the probability that $T \geq t$, i.e.,

$$R(t) = P(T \geq t) \tag{3.1.1}$$

$R(t)$ is also called *survival function* at time t. This defines a failure distribution of the component or a system given by its cumulative distribution function (c.d.f. >) as

$$F(t) = P(T < t). \tag{3.1.2}$$

If the failure distribution $F(t)$ is absolutely continuous possessing continuous derivative at every point, then

$$\frac{d}{dt}F(t) = f(t) \tag{3.1.3}$$

is called the *density function* of the failure system.

Intensity function of a failure system is the probability of failure in an infinitesimal interval in time δt as $\delta t \to 0$ (or orderliness condition). Under absence of orderliness condition one can define intensity function as

$$\lim_{t \to 0} \frac{1 - p_0(t)}{t},$$

where $p_0(t)$ represents the probability of no failure up to time t; and if we assume the probability of more than one failure between $(t, t+\delta t)$ is of the order zero as $\delta t \to 0$ (or orderless condition), then

$$\begin{aligned} & p_0(t + \delta t) = p_0(t)[1 - i(t)\delta t + 0(\delta t)] \\ \Rightarrow \quad & \lim_{\delta t \to 0} \frac{p_0(t + \delta t) - p_0(t)}{\delta t} = -i(t)p_0(t), \\ \text{i.e.,} \quad & p_0'(t) = -i(t)p_0(t). \end{aligned} \tag{3.1.4}$$

Hazard rate is defined to be the conditional probability of a failure between $(t, t+\delta t)$, given that there is no failure up to time t, as $\delta t \to 0$.

Hazard rate

$$h(t) = \frac{f(t)}{1-F(t)} = \frac{f(t)}{R(t)} \tag{3.1.5}$$

$$\Rightarrow \int_0^t h(\tau)\, d\tau = \int_0^t \frac{f(\tau)\, d\tau}{1-F(\tau)}$$

$$\Rightarrow \int_0^t h(\tau)\, d\tau = -\log_e(1-F(t)),$$

we have $F(0)=0$, $F(\infty)=1$, or, $R(0)=1$, $R(\infty)=0$ for a non-negative random variable.

$$\Rightarrow -\int_0^t h(\tau)\, d\tau = \log_e R(t)$$

$$\Rightarrow R(t) = e^{-\int_0^t h(\tau)\, d\tau}. \tag{3.1.6}$$

Example 1 *Show that for a negative exponential distribution*

$$f(t) = \lambda e^{-\lambda t};\ 0<\lambda<\infty,\ 0\le t<\infty;$$

intensity function $i(t)$ and hazard function $h(t)$ are the same.

Solution: We have probability of no failure up to t

$$= \lambda \int_t^\infty e^{-\lambda\tau}\, d\tau = e^{-\lambda t} = p_0(t).$$

$$p_0(t+\delta t) = p_0(t)[1 - i(t)\delta t + 0(\delta t)],$$

where

$$\lim_{t\to 0} \frac{1-p_0(t)}{t} = i(t)$$

$$\Rightarrow \lim_{t\to 0} \frac{1-e^{-\lambda t}}{t} = i(t) = \lim_{t\to 0} \frac{\lambda e^{-\lambda t}}{1} = \lambda$$

$$\Rightarrow i(t) = \lambda.$$

On the other hand,

$$R(t) = p_0(t) = e^{-\lambda t}$$

$$F(t) = 1 - e^{-\lambda t}$$

$$f(t) = \frac{d}{dt}(1-e^{-\lambda t}) = \lambda e^{-\lambda t}$$

$$h(t) = \frac{\lambda e^{-\lambda t}}{e^{-\lambda t}} = \lambda.$$

Hence,

$$i(t) = h(t) = \lambda$$

for a negative exponential distribution.

Remarks

In general $i(t) \neq h(t)$. The above equality is because of memoryless property of the negative exponential distribution $f(t) = \lambda e^{-\lambda t}$ for which $h(t)$ at any time t is always λ independent of t. We *say,* under this case the system forgets its age. In case of negative exponential distribution

$$p_0(t) = R(t) = e^{-\lambda t} \text{ and } \lim_{t \to 0} \frac{1 - p_0(t)}{t} = \lambda = h(t).$$

Example 2 *For Weibull distribution*

$$f(t) = \lambda e^{-\lambda_i^{\alpha}/\alpha \cdot t^{\alpha-1}};\ 0 < \lambda < \infty,$$

$t \geq 0$, $\alpha > 0$, or, < 0, $h(t) = i(t)$ under orderliness condition.

Solution: We have

$$F(t) = 1 - e^{-\lambda t^{\alpha}/\alpha},$$

$$R(t) = e^{-\lambda t^{\alpha}/\alpha},$$

$$h(t) = \frac{f(t)}{R(t)} = \frac{\lambda e^{-\frac{\lambda t^{\alpha}}{\alpha}} t^{\alpha-1}}{e^{-\frac{\lambda t^{\alpha}}{\alpha}}} = \lambda t^{\alpha-1},$$

a time dependent hazard rate.

If $\alpha > 1$, the hazard rate is increasing;
$\alpha < 1$, the hazard rate is decreasing;
$\alpha = 1$, the hazard rate is constant.

Now

$$p_0(t + \delta t) = p_0(t)(1 - i(t)\delta t) + 0(\delta t) = \lim_{\delta t \to 0} \frac{e^{-\frac{\lambda(t+\delta t)^{\alpha}}{\alpha}} - e^{-\frac{\lambda t^{\alpha}}{\alpha}}}{e^{-\frac{\lambda t}{\alpha}} \delta t} = -i(t)$$

$$\Rightarrow \quad \frac{p_0'(t)}{p_0(t)} = -i(t)$$

$$\Rightarrow \quad \frac{-e^{-\frac{\lambda t^{\alpha}}{\alpha}} \left(\frac{\lambda}{\alpha}\right) \alpha t^{\alpha-1}}{e^{-\frac{\lambda t^{\alpha}}{\alpha}}} = -i(t)$$

$$\Rightarrow \quad i(t) = \lambda t^{\alpha-1} = h(t).$$

Remarks

If orderliness condition does not hold, then $h(t)$ and $i(t)$ need not be the same. Under orderliness condition $h(t) = i(t)$. But we have seen that for a distribution with Loss of Memory Property (LMP) even under non-orderliness condition $h(t) = i(t)$. However, in general $h(t) \neq i(t)$.

Example 3 *Bath tub type of Hazard function.*

Given

$$\begin{aligned} h(t) &= \lambda_1 & t_0 = 0 &\leq t < t_1 \\ &= \lambda_2 & t_1 &\leq t < t_2 \\ &= \lambda_3 & t_2 &\leq t < t_3 \\ &\vdots & &\vdots \\ &= \lambda_{k-1} & t_{k-2} &\leq t < t_{k-1} \\ &= \lambda_k & t &\geq t_{k-1} \end{aligned}$$

Obtain $R(t)$ and $f(t)$.

Solution: We have

$$\begin{aligned} R(t) &= e^{-\lambda t};\ 0 \leq t < t_1 \\ &= e^{-\lambda_1 t_1} e^{-\lambda_2 (t-t_1)};\ t_1 \leq t < t_2 \\ &= e^{-\lambda_1 t_1} e^{-\lambda_2 t_2} e^{-\lambda_3 (t-t_2)};\ t_2 \leq t < t_3 \\ &\vdots \qquad \vdots \\ &= e^{-\lambda_1 t_1} e^{-\lambda_2 (t_2-t_1)} \ldots e^{-\lambda_{k-1}(t_{k-1}-t_{k-2})} e^{-\lambda_k (t-t_{k-1})};\ t \geq t_{k-1}. \end{aligned}$$

The density function

$$f(t) = \frac{d}{dt}[1 - R(t)]$$

accordingly is given by

$$\begin{aligned} f(t) &= \lambda_1 e^{-\lambda_1 t};\ 0 \leq t < t_1 \\ &= e^{-\lambda_1 t_1} \lambda_2 e^{-\lambda_2 (t-t_1)};\ t_1 \leq t < t_2 \\ &= e^{-\lambda_1 t_1} e^{-\lambda_2 (t_2-t_1)} \lambda_3 e^{-\lambda_3 (t-t_2)};\ t_2 \leq t < t_3. \end{aligned}$$

Proceeding in this way

$$f(t) = e^{-\lambda_1 t_1} \cdot e^{-\lambda_2 (t_2-t_1)}, \ldots, \lambda_k e^{-\lambda_k (t-t_{k-1})};\ t \geq t_{k-1}.$$

3.2 Distribution of Longevity and Its Moments

Let $f(t)$ be the density function of a failure distribution. The random variable T representing longevity is non-negative.

$$\begin{aligned} E(T^n) &= \int_0^\infty t^n f(t)\, dt \\ E(T^n) &= \int_0^\infty t^n \frac{dF(t)}{dt}\, dt = -\int_0^\infty t^n \frac{dR(t)}{dt}\, dt. \end{aligned}$$

Integrating by parts we have

$$E(T^n) = -\Big[t^n R(t)\Big]_0^\infty + n \int_0^\infty t^{n-1} R(t)\, dt.$$

We have $R(\infty) = 0$ and $R(0) = 1$

$$\Rightarrow E(T^n) = n\int_0^\infty t^{n-1}R(t)\,dt. \tag{3.2.1}$$

In particular, for $n = 1$

$$E(T) = \int_0^\infty R(t)\,dt. \tag{3.2.2}$$

For $n = 2$

$$E(T^2) = 2\int_0^\infty tR(t)\,dt \tag{3.2.3}$$

$$\Rightarrow \operatorname{var}(T) = 2\int_0^\infty tR(t)\,dt - \left[\int_0^\infty R(t)\,dt\right]^2. \tag{3.2.4}$$

By the same reasoning $E[f(T)]$, where $f(T)$ is a measurable function of absolutely continuous non-negative random variables T,

$$E[f(T)] = \int_0^\infty f'(t)R(t)\,dt + f(0) \tag{3.2.5}$$

$$\because\ E[f(T)] = -\int_0^\infty f(t)\frac{dR(t)}{dt}\,dt$$

$$= -\Big[f(t)R(t)\Big]_0^\infty + \int_0^\infty f'(t)R(t)\,dt$$

$$\text{i.e.,}\quad = f(0) + \int_0^\infty f'(t)R(t)\,dt. \tag{3.2.6}$$

If $f(0) = 0$, then (3.2.6) is reduced to

$$E[f(t)] = \int_0^\infty f'(t)R(t)\,dt. \tag{3.2.7}$$

3.3 Classification of Systems

A system is defined as the combination of several units or components. If the system fails on the failure of any of the units it is called a *Series system.*

On the other hand, if the system survives till the survival of the last component, it is called a *Parallel system.* In the human bio-systems, Kidney system, Respiratory system, Audio and Visual systems are all parallel systems. On the other hand, a radio ceases to function on the failure of a single valve or a condenser; or a refrigerator fails to work on the failure of a particular component, *say,* expansion valve, etc. These are all the examples of series systems.

Besides the above two systems, in some situations it may happen that a fixed number of identical components (*say,* m out of n; $m < n$) should be in working condition in order that system works. Such a system is neither parallel nor series. They may be named as *Mixed systems.* A well-known example of a mixed system is the *four engine Aircraft system,* where the system will work if two out of four engines are in working condition, sometimes subject to more restrictive condition, viz., at least one engine on either side should work.

An altogether different basis of classifying a system is based on *Maintenance Condition.* That is, on the failure of a component, if there is a provision of servicing the same by the user of the system, then the system is called *Maintained system.* On the other hand, if the system cannot be repaired by the user, it is called a *Non-maintained system.*

In a non-maintained system, immediate replacement of a component on its failure is a bare necessity to avoid unnecessary hampering or delay in the production process. Sometimes a duplicate component is kept as spare and is replaced in the main system on failure of the original one. It is called a *standby* and the system is known as a *Standby system.*

If the spare component is kept idle till it is put in the main system on failure of the original one then the standby is known as *Cold standby.* On the other hand, if the spare component is used in a less hazardous system and used in the main system when required, then it is called a *Warm standby.*

3.3.1 A replacement model

Let the time of failure of identical components have a common cumulative distribution function $F(t)$ and hazard rate $\lambda(t)$. Let $P_n(t)$ denote the probability of n replacements in $(0, t)$. We assume the condition of orderliness of the process (i.e., in an infinitesimal interval of time there cannot be more than one failure). Then

$$P_n(t+\delta t) = P_{n-1}(t)[\lambda(t)\delta t + 0(\delta t)] + P_n(t)[1 - \{\lambda(t)\delta t + 0(\delta t)\}] \tag{3.3.1}$$

$$\lim_{\delta t \to 0} \frac{P_n(t+\delta t) - P_n(t)}{\delta t} = P_{n-1}(t)\lambda(t) - P_n(t)\lambda(t) + \lim_{\delta t \to 0} \frac{0(\delta t)}{\delta t}$$

$$\frac{dP_n(t)}{dt} = [P_{n-1}(t) - P_n(t)]\lambda(t). \tag{3.3.2}$$

Also, for $n = 0$,

$$P_0(t+\delta t) = P_0(t)[1 - \{\lambda(t)\delta t + 0(\delta t)\}]. \tag{3.3.3}$$

$$\lim_{\delta t \to 0} \frac{P_0(t+\delta t) - P_0(t)}{\delta t} = -\lambda(t)P_0(t)$$

$$\Rightarrow \quad P_0'(t) = -\lambda(t)P_0(t).$$

Integrating

$$\log P_0(t) = -\int_0^\infty \lambda(\tau)\, d\tau.$$

$$P_0(t) = e^{-\int_0^t \lambda(\tau)\, d\tau} = e^{-\wedge(t)}, \ (say),$$

where

$$\wedge(t) = \int_0^t \lambda(\tau)\, d\tau.$$

For $n = 1$.

$$\frac{dP_1(t)}{dt} = [P_0(t) - P_1(t)]\lambda(t)$$

$$\Rightarrow \quad \frac{dP_1(t)}{dt} + \lambda(t)P_1(t) = \lambda(t)P_0(t).$$

Integrating factor $= e^{\int_0^t \lambda(\tau)\, d\tau} = e^{\wedge(t)}$

$$e^{\wedge(t)}\frac{dP_1(t)}{dt} + \lambda(t)e^{\wedge(t)}\, P_1(t) = \lambda(t)P_0(t)e^{\wedge(t)}.$$

$$\frac{d}{dt}\left[P_1(t)e^{\wedge(t)}\right] = \lambda(t); \text{ since } P_0(t) = e^{-\wedge(t)}$$

$$\left[P_1(\tau)e^{\wedge(\tau)}\right]_0^t = \int_0^t \lambda(\tau)\, d\tau;\ P_1(0) = 0,\ \lambda(0) = 0$$

$$\Rightarrow \quad P_1(t) = e^{-\wedge(t)} \wedge (t) = e^{-\int_0^t \lambda(\tau)\, d\tau} \int_0^t \frac{\lambda(\tau)}{1!}\, d\tau. \tag{3.3.4}$$

Next let us write

$$P_n(t) = e^{-\wedge(t)}\frac{[f(t)]^n}{n!} \tag{3.3.5}$$

as a trial solution of (3.3.2)

$$\frac{dP_n(t)}{dt} = [P_{n-1}(t) - P_n(t)]\, \lambda(t);\ n \neq 0 \tag{3.3.6}$$

$$\Rightarrow \frac{dP_n(t)}{dt} = e^{-\wedge(t)} n \frac{[f(t)]^{n-1}}{n!} f'(t) - \wedge(t) e^{-\wedge(t)} \frac{[f(t)]^n}{n!}. \tag{3.3.7}$$

Substituting (3.3.5) in (3.3.6)

$$\Rightarrow \frac{dP_n(t)}{dt} = \lambda(t)e^{-\wedge(t)}\left[\frac{[f(t)]^{n-1}}{(n-1)!} - \frac{[f(t)]^n}{n!}\right]. \tag{3.3.8}$$

Comparing (3.3.7) and (3.3.8), we get

$$\Rightarrow \quad f'(t) = \lambda(t)$$

$$\Rightarrow \quad f(t) = \int_0^t \lambda(\tau)\, d\tau. \tag{3.3.9}$$

Hence putting (3.3.9) against $f(t)$ in (3.3.5)

$$\Rightarrow \quad P_n(t) = e^{-\wedge(t)} \frac{\left(\int_0^t \lambda(\tau)\, d\tau\right)^n}{n!}$$

$$\Rightarrow \quad P_n(t) = e^{-\int_0^t \lambda(\tau)\, d\tau} \frac{\left(\int_0^t \lambda(\tau)\, d\tau\right)^n}{n!}. \tag{3.3.10}$$

Example 4 *For the replacement model*

$$P[X(t) = n \mid \lambda(t)] = e^{-\int_0^t \lambda(\tau)\, d\tau} \frac{\left(\int_0^t \lambda(\tau)\, d\tau\right)^n}{n!},$$

when $\lambda(t) = \lambda t^{\alpha-1}$, obtain the waiting time distribution for the first failure. Obtain the result in a particular case when $\lambda(t) = \lambda$ (i.e., the intensity is time independent).

Solution:

$$P[X(t)=0] = e^{-\int_0^t \lambda(\tau)\,d\tau} = e^{-\frac{\lambda t^\alpha}{\alpha}}$$
$$P[X(t)=0] = R(t) = P[T_1 > t] = e^{-\frac{\lambda t^\alpha}{\alpha}},$$

where T_1 is the time for first failure.

$$\Rightarrow F(t) = P[T_1 \le t] = 1 - e^{-\frac{\lambda t^\alpha}{\alpha}}$$
$$f(t) = \frac{dF(t)}{dt} = e^{-\frac{\lambda t^\alpha}{\alpha}} \frac{\lambda \alpha t^{\alpha-1}}{\alpha} = \lambda e^{-\frac{\lambda t^\alpha}{\alpha}} t^{\alpha-1}.$$

This is a Weibull distribution. For $\alpha = 1$, the distribution is $\lambda e^{-\lambda t}$ which is negative exponential.

Example 5 *If the failure distribution conforms to a Poisson process, then obtain the waiting time distribution of nth failure* $(n > 1)$.

Solution: We have

$$P[n \text{ failures in } (0,t)] = P[X(t)=n] = e^{-\lambda t}\frac{(\lambda t)^n}{n!} = P_n(t)$$
$$P_0(t) = P[X(t)=0] = e^{-\lambda t}$$
$$P_1(t) = P[X(t)=1] = \lambda e^{-\lambda t}$$
$$\Rightarrow \quad R_1(t) = P[T_1 > t] = e^{-\lambda t},$$

where T_1 is the time of first failure.

$$F(t) = P[T_1 \le t] = 1 - e^{-\lambda t}$$
$$f_1(t) = \frac{d}{dt}(1 - e^{-\lambda t}) = \lambda e^{-\lambda t}$$
$$P[T_2 > t] = P_0(t) + P_1(t) = e^{-\lambda t} + \lambda t e^{-\lambda t},$$

where T_2 is the time for the second failure.

$$P[T_2 \le t] = 1 - (e^{-\lambda t} + \lambda t^{-\lambda t}).$$

Therefore, the waiting time distribution for the second failure is given by

$$\begin{aligned} f_2(t) &= \frac{d}{dt}[1 - (e^{-\lambda t} + \lambda t e^{-\lambda t})] \\ &= \lambda e^{-\lambda t} + \lambda^2 t e^{-\lambda t} - \lambda e^{-\lambda t} \\ &= \lambda^2 t e^{-\lambda t} \\ \text{or, } f_2(t) &= \frac{\lambda^2 e^{-\lambda t} t^{2-1}}{\Gamma(2)}. \end{aligned}$$

This is the probability density function of Gamma distribution with parameters 2 and λ. Precisely, proceeding in this way the waiting time distribution of the third failure is given by

$$f_3(t) = \frac{d}{dt}\left[1 - \left\{(e^{-\lambda t} + \lambda t e^{-\lambda t}) + \frac{(\lambda t)^2}{2!}e^{-\lambda t}\right\}\right].$$

We get

$$f_3(t) = \frac{\lambda^3 e^{-\lambda t} t^{3-1}}{\Gamma(3)} \text{ and } f_n(t) = \frac{\lambda^n e^{-\lambda t} t^{n-1}}{\Gamma(n)}.$$

Example 6 (A problem of parallel system: A kidney problem—Gross and Clerk, 1975) *This is in respect of a disease called* Renal Calculi *which inevitably affects kidney system. Initially either of the kidneys may fail, say with common intensity* λ_0. *If one of the kidneys ceases functioning, the intensity of failure of second kidney is increased from* λ_0 *to* λ_1 $(\lambda_1 > 2\lambda_0)$. *Obtain the Reliability of the kidney system and its expected longevity.*

Solution: With the usual notations, the Kolmogorov equations are

$$\begin{aligned} p_0(t+\delta t) &= p_0(t)[1-\{2\lambda_0\delta t + 0(\delta t)\}] && \text{(i)}\\ p_1(t+\delta t) &= p_0(t)[2\lambda_0\delta t + 0(\delta t)] + p_1(t)[1-\{\lambda_1\delta t + 0(\delta t)\}] && \text{(ii)}\\ \text{and } p_2(t+\delta t) &= p_1(t)[\{\lambda_1\delta t + 0(\delta t)\}] + p_2(t). && \text{(iii)} \end{aligned}$$

From (i) we have,

$$\lim_{\delta t\to 0} \frac{p_0(t+\delta t)-p_0(t)}{\delta t} = -2\lambda_0 p_0(t)$$

$$\Rightarrow \quad p_0'(t) = -2\lambda_0 p_0(t).$$

Integrating, $\log p_0(t) = -2\lambda_0 t$ since $p_0(0) = 1$, $\log p_0(0) = 0$.

$$\Rightarrow p_0(t) = e^{-2\lambda_0 t}. \tag{iv}$$

Next from (ii),

$$\begin{aligned} \lim_{\delta t\to 0} \frac{p_1(t+\delta t)-p_1(t)}{\delta t} &= -\lambda_1 p_1(t) + 2\lambda_0 p_0(t) + \lim_{\delta t\to 0} \frac{0(\delta t)}{\delta t}\\ \Rightarrow p_1'(t) &= -\lambda_1 p_1(t) + 2\lambda_0 p_0(t). \end{aligned}$$

$$\begin{aligned} \frac{dp_1(t)}{dt} + \lambda_1 p_1(t) &= 2\lambda_0 e^{-2\lambda_0 t}, \text{ substituting from } (v)\\ \Rightarrow e^{\lambda_1 t}\frac{dp_1(t)}{dt} + \lambda_1 e^{\lambda_1 t} p_1(t) &= 2\lambda_0 e^{-2\lambda_0 t} e^{\lambda_1 t}\\ \Rightarrow \frac{d}{dt}\left(p_1(t)e^{\lambda_1 t}\right) &= 2\lambda_0 e^{t(\lambda_1 - 2\lambda_0)}. \end{aligned}$$

$$\begin{aligned} p_1(t)e^{\lambda_1 t} &= 2\lambda_0 \int e^{t(\lambda_1-2\lambda_0)}\,dt + c \text{ (constant)}\\ p_1(t)e^{\lambda_1 t} &= \frac{2\lambda_0 e^{t(\lambda_1-2\lambda_0)}}{\lambda_1 - 2\lambda_0} + c. \end{aligned}$$

At $t = 0$,

$$p_1(0) = 0 \;\Rightarrow\; 0 = \frac{2\lambda_0}{\lambda_1 - 2\lambda_0} + c$$

$$\Rightarrow\; c = \frac{2\lambda_0}{2\lambda_0 - \lambda_1};\; \lambda_1 > 2\lambda_0$$

$$\Rightarrow p_1(t) = \frac{2\lambda_0 e^{-2\lambda_0 t}}{\lambda_1 - 2\lambda_0} - \frac{2\lambda_0}{\lambda_1 - 2\lambda_0}e^{-\lambda_1 t} = \frac{2\lambda_0}{\lambda_1 - 2\lambda_0}\left(e^{-2\lambda_0 t} - e^{-\lambda_1 t}\right). \tag{v}$$

$$\therefore\ R(t) = p_0(t) + p_1(t) = \frac{2\lambda_0}{\lambda_1 - 2\lambda_0}(e^{-2\lambda_0 t} - e^{-\lambda_1 t}) + e^{-2\lambda_0 t}. \tag{vi}$$

$$\begin{aligned}\text{Expected longevity} = E(T) &= \int_0^\infty \left[e^{-2\lambda_0 t} + \frac{2\lambda_0}{\lambda_1 - 2\lambda_0}(e^{-2\lambda_0 t} - e^{-\lambda_1 t})\right] dt \\ &= \frac{e^{-2\lambda_0 t}}{-2\lambda_0}\Big|_0^\infty + \frac{2\lambda_0}{\lambda_1 - 2\lambda_0}\left[\frac{e^{-2\lambda_0 t}}{-2\lambda_0} - \frac{e^{-\lambda_1 t}}{-\lambda_1}\right]_0^\infty \\ &= \frac{1}{-2\lambda_0} + \frac{2\lambda_0}{\lambda_1 - 2\lambda_0}\left[\frac{1}{2\lambda_0} - \frac{1}{\lambda_1}\right] > 0. \end{aligned} \tag{vii}$$

since

$$\frac{1}{2\lambda_0} > \frac{1}{\lambda_1} \text{ as } \lambda_1 > 2\lambda_0.$$

It can be shown that

$$f(t) = \frac{dF(t)}{dt} = -\frac{dR(t)}{dt} = \frac{2\lambda_0\lambda_1}{\lambda_1 - 2\lambda_0}(e^{-2\lambda_0 t} - e^{-\lambda_1 t});$$

for $t \geq 0$, $\lambda_1 > 2\lambda_0$, on simplification.

$$h(t) = \frac{f(t)}{R(t)} = \frac{2\lambda_0\lambda_1}{(\lambda_1 - 2\lambda_0)\left[1 - e^{-(\lambda_1 - 2\lambda_0)t}\right]^{-1} + 2\lambda_0},$$

$h(t)$ is an increasing function of t for $\lambda_1 \neq 2\lambda_0$; $h(0) = 0$, $h(\infty) = 2\lambda_0$.

Example 7 (A problem of series system) *In a k-component series system let the ith component has a hazard rate $\lambda_i(t)$; $i = 1, 2, \ldots, k$. Obtain the reliability and the waiting time density of the system.*

Solution: We have

$$R(t) = P[T > t] = \prod_{i=1}^{k} e^{-\int_0^t \lambda_i(\tau)\,d\tau} = \prod_{i=1}^{k} e^{-\Lambda_i(t)},$$

where

$$\int_0^t \lambda_i(\tau)\,d\tau = \Lambda_i(t) = \exp -\sum_{i=1}^{k} \Lambda_i(t).$$

Cumulative distribution function

$$F(t) = 1 - \exp -\sum_{i=1}^{k} \Lambda_i(t).$$

Density function of the failure system

$$= \frac{d}{dt}F(t) = \sum_{i=1}^{k} \lambda_i(t) \exp -\sum_{i=1}^{k} \Lambda_i(t).$$

Example 8 (An example of the mixed system) *We have an m component system. Assuming that all components have identical hazard rates $\lambda(t)$ and the system may work if at least $k(k < m)$ are in working condition, obtain the reliability and the failure density of the system.*

Solution:

$$\begin{aligned}
R(t) &= \sum_{i=k}^{m} P(i \text{ number of components do not fail up to } t) \\
&= \sum_{i=k}^{m} \binom{m}{i} \left(e^{-\int_0^t \lambda(\tau)\,d\tau}\right)^i \left(1 - e^{-\int_0^t \lambda(\tau)\,d\tau}\right)^{m-i}. \\
\therefore\ f(t) &= \frac{d}{dt}(1 - R(t)) \\
&= \sum_{i=k}^{m} \binom{m}{i} i \left(e^{-\lambda(t)}\right)^{t-1} \left(1 - e^{-\Lambda(t)}\right)^{m-1} + \binom{m}{i} \left(e^{-\Lambda(t)}\right)^i (m-i) \\
&\quad \left(i - e^{-\Lambda(t)}\right)^{m-i-1} \lambda(t).
\end{aligned}$$

Example 9 (Another example of mixed system) *There are four engines in an Aircraft. Each engine has a time independent failure rate. Obtain the reliability of the Aircraft, given that (1) at least two engines should work, (2) at least two engines, one each on either side should work. Also obtain the longevity of the system on either of the cases.*

Solution: Assuming that the common failure intensity or the hazard rate is λ, we have the following Kolmogorov equations under (1):

$$\begin{aligned}
p_0(t+\delta t) &= p_0(t)[1 - 4\lambda\delta t + 0(\delta t)] && \text{(i)} \\
p_1(t+\delta t) &= p_0(t)[4\lambda\delta t + 0(\delta t)] + p_1(t)[1 - 3\lambda\delta t + 0(\delta t)] && \text{(ii)} \\
p_2(t+\delta t) &= p_1(t)[3\lambda\delta t + 0(\delta t)] + p_2(t)[1 - 2\lambda\delta t + 0(\delta t)]. && \text{(iii)}
\end{aligned}$$

On the other hand, Kolmogorov equations under (2) will keep (*i*) and (*ii*) to be the same but (*iii*) will be changed to

$$p_2(t+\delta t) = p_1(t)[2\lambda\delta t + 0(\delta t)] + p_2(t)[1 - \{2\lambda\delta t + 0(\delta t)\}]. \qquad \text{(iv)}$$

As usual (*i*), (*ii*), (*iii*) $\Rightarrow$

$$\begin{aligned}
p_0'(t) &= -4\lambda p_0(t) && \text{(v)} \\
p_1'(t) &= -3\lambda p_1(t) + 4\lambda p_0(t) && \text{(vi)} \\
p_2'(t) &= -2\lambda p_2(t) + 3\lambda p_1(t). && \text{(vii)}
\end{aligned}$$

Let

$$\begin{aligned}
L(p_n(t)) &= \int_0^\infty e^{-st} p_n(t)\,dt \\
&= \text{Laplace's transformation of } p_n(t),\ n = 0, 1, 2, \ldots \\
L(p_0'(t)) &= \int_0^\infty e^{-st} p_0'(t)\,dt \\
&= e^{-st} p_0(t)\Big|_0^\infty + s\int_0^\infty e^{-st} p_0(t)\,dt \\
&= -1 + s\int_0^\infty e^{-st} p_0(t)\,dt, \text{ since } p_0(0) = 1 \\
\Rightarrow L(p_0'(t)) &= -1 + sL(p_0(t)), && (3.3.11)
\end{aligned}$$

whereas

$$\begin{aligned} L(p_n'(t)) &= \int_0^\infty e^{-st} p_n'(t)\,dt, \text{ for } n = 1, 2, 3, \ldots \\ &= e^{-st} p_n(t)\Big|_0^\infty + s\int_0^\infty e^{-st} p_n(t)\,dt \\ &= sL(p_n(t)), \end{aligned} \tag{3.3.12}$$

From (v),

$$\begin{aligned} &\Rightarrow -1 + sL(p_0(t)) = -4\lambda L(p_0(t)) \\ &\Rightarrow (s+4\lambda)L(p_0(t)) = 1 \\ &\Rightarrow L(p_0(t)) = \frac{1}{s+4\lambda} \\ &\Rightarrow p_0(t) = e^{-4\lambda t}. \end{aligned} \tag{3.3.13}$$

Similarly from (vi),

$$\begin{aligned} L(p_1'(t)) = SL(p_1(t)) &= -3\lambda L(p_1(t)) + 4\lambda L(p_0(t)) \\ \Rightarrow (s+3\lambda)L(p_1(t)) &= 4\lambda L(p_0(t)) = \frac{4\lambda}{s+4\lambda} \text{ [from (3.3.13)]} \\ \Rightarrow L(p_1(t)) &= \frac{4\lambda}{(s+3\lambda)(s+4\lambda)} \\ \Rightarrow L(p_1(t)) &= \frac{4\lambda}{\lambda}\left[\frac{1}{(s+3\lambda)} - \frac{1}{(s+4\lambda)}\right] \\ L(p_1(t)) &= 4\left[\frac{1}{(s+3\lambda)} - \frac{1}{(s+4\lambda)}\right]. \\ \Rightarrow p_1(t) &= 4(e^{-3\lambda t} - e^{-4\lambda t}). \end{aligned} \tag{3.3.14}$$

Finally from (vii),

$$\begin{aligned} sL(p_2(t)) &= -2\lambda L(p_2(t)) + 3\lambda L(p_1(t)) \\ \Rightarrow (s+2\lambda)L(p_2(t)) &= 3\lambda L(p_1(t)) \\ \Rightarrow L(p_2(t)) &= \frac{3\lambda L(p_1(t))}{(s+2\lambda)} = \frac{3\lambda}{(s+2\lambda)}4\left[\frac{1}{(s+3\lambda)} - \frac{1}{(s+4\lambda)}\right] \\ &= \left[\frac{12}{(s+2\lambda)} - \frac{12}{(s+3\lambda)}\right] - \left[\frac{6}{(s+2\lambda)} - \frac{6}{(s+4\lambda)}\right] \text{ [from (3.3.14)]} \\ p_2(t) &= 12e^{-2\lambda t} - 12e^{-3\lambda t} - 6e{-2\lambda t} + 6e^{-4\lambda t} \\ &= 6e^{-2\lambda t} - 12e^{-3\lambda t} + 6e^{-4\lambda t}. \end{aligned} \tag{3.3.15}$$

The reliability of the Aircraft under (1) is, therefore,

$$\begin{aligned} R(t) &= p_0(t) + p_1(t) + p_2(t) \\ &= 3e^{-4\lambda t} - 8e^{-3\lambda t} + 6e^{-2\lambda t}. \end{aligned} \tag{3.3.16}$$

It can be checked to see $R(0) = 1$, $R(\infty) = 0$.

$$\text{Expected longevity} = \int_0^{\alpha} R(t)\,dt$$

$$\begin{aligned} E(T) &= 3\int_0^{\infty} e^{-4\lambda t}\,dt - 8\int_0^{\infty} e^{-3\lambda t}\,dt + 6\int_0^{\infty} e^{-2\lambda t}\,dt \\ &= \frac{3}{4\lambda} - \frac{8}{3\lambda} + \frac{6}{2\lambda} = \frac{13}{12\lambda} > 0. \end{aligned} \tag{3.3.17}$$

On the other hand, under (2), (*vii*)

$$\begin{aligned} \Rightarrow p_2'(t) &= -2\lambda p_2(t) + 2\lambda p_1(t) \\ \Rightarrow sL(p_2(t)) &= -2\lambda L(p_2(t)) + 2\lambda L(p_1(t)) \\ \Rightarrow (s+2\lambda)L(p_2(t)) &= 2\lambda L(p_1(t)) \end{aligned} \tag{3.3.18}$$

$$\begin{aligned} \Rightarrow L(p_2(t)) &= \frac{2\lambda}{(s+2\lambda)}L(p_1(t)) = \frac{2\lambda}{(s+2\lambda)}4\left[\frac{1}{(s+3\lambda)} - \frac{1}{(s+4\lambda)}\right], \text{ for (3.24)} \\ &= 8\left[\frac{1}{(s+2\lambda)} - \frac{1}{(s+3\lambda)}\right] - 4\left[\frac{1}{(s+2\lambda)} - \frac{1}{(s-4\lambda)}\right] \end{aligned}$$

$$\begin{aligned} p_2(t) &= 8(e^{-2\lambda t} - e^{-3\lambda t}) - 4(e^{-2\lambda t} - e^{-4\lambda t}) \\ &= 4e^{-2\lambda t} - 8e^{-3\lambda t} + 4e^{-4\lambda t}. \end{aligned} \tag{3.3.19}$$

Therefore, the reliability of the Aircraft under (2) is given by

$$\begin{aligned} R(t) &= p_0(t) + p_1(t) + p_2(t) \\ &= e^{-4\lambda t} + 4(e^{3\lambda t} - e^{-4\lambda t}) + 4e^{-2\lambda t} - 8e^{-3\lambda t} + 4e^{-4\lambda t} \\ &= e^{-4\lambda t} - 4e^{-3\lambda t} + 4e^{-2\lambda t}. \end{aligned} \tag{3.3.20}$$

Expected longevity

$$\begin{aligned} E(T) &= \int_0^{\infty} e^{-4\lambda t}\,dt - 4\int_0^{\infty} e^{-3\lambda t}\,dt + 4\int_0^{\infty} e^{-2\lambda t}\,dt \\ &= \frac{1}{4\lambda} - \frac{4}{3\lambda} + \frac{4}{2\lambda} = \frac{3-16+24}{12\lambda} = \frac{11}{12\lambda}. \end{aligned} \tag{3.3.21}$$

On comparison of (3.3.17) and (3.3.21) it follows that expected longevity is reduced by

$$\left(\frac{13}{12\lambda} - \frac{11}{12\lambda}\right) = \frac{1}{6\lambda},$$

because of imposition of further restriction.

Example 10 (On unit system with one Warm Standby of Non-Maintained System) *There is a one unit system with λ_1 as 'on line' hazard rate. The standby is being used in a different line (till the original gets out of order) with lower hazard rate $\lambda_2(< \lambda_1)$. Obtain the longevity of the system.*

Solution: We have the Kolmogorov equations given by

$$\begin{aligned} p_0(t+\delta t) &= p_0(t)[1-\lambda_1\delta t-\lambda_2\delta t+0(\delta t)] \\ p_1(t+\delta t) &= p_0(t)[\lambda_1\delta t+\lambda_2\delta t+0(\delta t)]+p_1(t)[1-\lambda_1\delta t+0(\delta t)]. \end{aligned}$$

Since the standby must be on the line if one component fails, its hazard rate becomes λ_1; otherwise if the standby fails the original remains on the line with hazard rate λ_1.

It is seen that,

$$\begin{aligned} R(t) &= \frac{\lambda_1+\lambda_2}{\lambda_2}e^{-\lambda_1 t}-\frac{\lambda_1}{\lambda_2}e^{-(\lambda_1+\lambda_2)t} \\ \text{and } L &= \frac{\lambda_1+\lambda_2}{\lambda_1\lambda_2}-\frac{\lambda_1}{\lambda_2(\lambda_1+\lambda_2)}, \end{aligned}$$

L = longevity of the system.

3.4 A Note on Availability

Availability can be thought of as Point Reliability, whereas Reliability in the proper sense of term means *Interval Reliability*. In the two state Markov (*vide* Section 3.5) chain a single equipment with no redundancy the point Reliability or Availability at t is (i.e., the probability of being available at t, given the same is available at o) denoted by $P_{00}\ (t)$,

$$\text{i.e., } p_{00}\ (t)=\frac{\mu}{\lambda+\mu}+\frac{\lambda}{\lambda+\mu}\left(1-e^{-(\lambda+\mu)t}\right)\quad [\textit{vide } (3.5.2)]$$

λ is the failure rate and μ is the restoration rate.

3.5 Maintained System: Single Equipment with no Redundancy—A Problem of Two-State Markov Chain in Continuous Time

Suppose, we have a problem of a single component system. The intensity of failure is λ and that of coming back to working condition is μ. Then the problem lies in finding out that the given system is in working condition/state of repairment what is the probability of the same in working condition/under repairment? Also to find out what fraction or proportion of time the system is in working condition or otherwise in $(0, t]$.

Let us denote the two states as $i = 0, 1$; '0' may be taken as the state of working condition and '1' otherwise. Let p_{ij} represent the transition probability of the system going to $j (j = 0, 1)$, given that it is initially in state $i(0, 1)$. Then we have the Kolmogorov equations,

$$p_{00}\ (t+\delta t)=p_{00}\ (t)[1-(\lambda\delta t+0(\delta t)]+p_{01}\ (t)[\mu\delta t+0(\delta t)] \qquad (3.5.1)$$

$$\Rightarrow \lim_{\delta t\to 0}\frac{p_{00}\ (t+\delta t)-p_{00}\ (t)}{\delta t}=-\lambda p_{00}\ (t)+\mu p_{01}\ (t)+\lim_{\delta t\to 0}\frac{0(\delta t)}{\delta t}$$

$$\Rightarrow p'_{00}\ (t)=-\lambda p_{00}\ (t)+\mu p_{01}\ (t)).$$

Taking Laplace transform on both sides

$$\Rightarrow -1+sL(p_{00}\ (t))=-\lambda L(p_{00}\ (t))+\mu L(1-p_{00}\ (t)),$$

since $p_{00}(t) + p_{01}(t) = 1$

$$\text{or, } (s+\lambda+\mu)L(p_{00}(t)) = \mu L(1) + 1 = \mu/s + 1$$

$$\begin{aligned}
\Rightarrow L(p_{00}(t)) &= \frac{\mu/s+1}{s+\lambda+\mu} = \frac{\mu}{s(s+\lambda+\mu)} + \frac{1}{s(s+\lambda+\mu)} \\
&= \left(\frac{1}{s} - \frac{1}{s+\lambda+\mu}\right)\frac{\mu}{\lambda+\mu} + \frac{1}{s+\lambda+\mu} \\
\Rightarrow p_{00}(t) &= \frac{\mu}{\lambda+\mu} - \frac{\mu}{\lambda+\mu}e^{-(\lambda+\mu)t} + e^{-(\lambda+\mu)t} \\
&= \frac{\mu}{\lambda+\mu} + \left(1 - \frac{\mu}{\lambda+\mu}\right)e^{-(\lambda+\mu)t} \\
&= \frac{\mu}{\lambda+\mu} + \frac{\lambda}{\lambda+\mu}e^{-(\lambda+\mu)t} \qquad (3.5.2) \\
p_{01}(t) = 1 - p_{00}(t) &= \frac{\lambda}{\lambda+\mu} - \frac{\lambda}{\lambda+\mu}e^{-(\lambda+\mu)t} \qquad (3.5.3)
\end{aligned}$$

$p_{10}(t)$ is obtained by replacing λ and μ and vice versa in $p_{01}(t)$, i.e.,

$$p_{10}(t) = \frac{\mu}{\lambda+\mu} - \frac{\mu}{\lambda+\mu}e^{-(\lambda+\mu)t} \qquad (3.5.4)$$

and finally

$$p_{11}(t) = 1 - p_{10}(t) = \frac{\lambda}{\lambda+\mu} + \frac{\mu}{\lambda+\mu}e^{-(\lambda+\mu)t}. \qquad (3.5.5)$$

Given that initially the system starts with '0' state, the average time in $(0,t)$ for which it remains in '0' (i.e:, *intensity*), $\mu_{00}(t)$ is obviously obtained from the following consideration. Let $Y(t)$ be the random variable and defined as

$$Y(t) = \begin{cases} 1 & \text{if } \{X(t) = 0 \mid X(0) = 0\} \\ 0 & \text{otherwise,} \end{cases}$$

where $Y(t) = 1$ is the event representing the stay of the system in state '0' at time t, given that it remained in '0' initially. Therefore,

$$E(Y(t)) = p_{00}(t)\cdot 1 + [1 - p_{00}(t)]\cdot 0 = p_{00}(t).$$

Hence the average time for which the system remains '0' in $(0,t)$, given that it remained in '0' initially, is

$$\begin{aligned}
\mu_{00}(t) &= \int_0^t p_{00}(\tau)\,d\tau = \frac{\mu}{\lambda+\mu}t + \left(\frac{\lambda}{\lambda+\mu}\int_0^t \left[e^{-(\lambda+\mu)\tau}\,d\tau\right]\right) \\
&= \frac{\mu}{\lambda+\mu}t + \frac{\lambda}{(\lambda+\mu)^2}\left[1 - e^{-(\lambda+\mu)t}\right] \text{ [from (3.5.2)]} \qquad (3.5.6)
\end{aligned}$$

Precisely with similar arguments

$$\mu_{01}(t) = t - \mu_{00}(t) = \int_0^t p_{01}(\tau)\,d\tau$$

$$= \frac{\lambda}{\lambda+\mu}t - \frac{\lambda}{(\lambda+\mu)^2}\left[1-e^{-(\lambda+\mu)t}\right] \text{ [from (3.5.3)]} \quad (3.5.7)$$

$$\mu_{10}(t) = \frac{\mu}{\lambda+\mu}t - \frac{\mu}{(\lambda+\mu)^2}\left[1-e^{-(\lambda+\mu)t}\right] \quad (3.5.8)$$

$$\mu_{11}(t) = \frac{\mu}{\lambda+\mu}t - \frac{\mu}{(\lambda+\mu)^2}\left[1-e^{-(\lambda+\mu)t}\right] \quad (3.5.9)$$

Remarks

One major application of (3.5.6) to (3.5.9) is to estimate the expected proportion of period in $(0, t)$, and it may be taken as a year, for which one can utilise the services of the system. *For example,* on a lead factory, the industrial workers are exposed to lead poisoning and they are allowed leave under Industrial Act. Nevertheless, this is a kind of leakage to the Management because workers are granted leave without work. Since such phenomena are quite frequent in Industrial concern, the Management may like to examine the extent of loss in production, etc. on account of such recurring casualties. This is one of the major applications in such practical situations. The probability distribution of the period for which the system remains incorporative has been given by Barlow and Hunter(1961). This is derived by specialising a general result due to Tackas(1957). It has been shown that if $I(t)$ is the time for which the system is operative

$$P[I(t) \le x] = e^{-\lambda(t-x)}\left(1 + \sqrt{\lambda\mu(t-x)}\int_0^x e^{-\mu y} y^{-1/2} I_1\left[2\sqrt{\lambda\mu(t-x)y}\right] dy\right), \quad (3.5.10)$$

where $I_1(x)$ is the Bessel function of order 1 for imaginary argument

$$I_1(x) = \sum_{j=0}^{\infty} \frac{(1/2x)^{2j+1}}{j!(j+1)!}. \quad (3.5.11)$$

Bessel function tables are readily available. For practical purposes asymptotic distribution of $I(t)$ is more useful.

$$I(t) \overset{a.d.}{\sim} N\left(\frac{\lambda\delta t}{\lambda+\mu}, \sqrt{\frac{2\lambda\mu t}{(\lambda+\mu)^3}}\right). \quad (3.5.12)$$

3.6 Maintained System with Hot Standby Redundancy (Redundant Unit Fails with the Same Rate)

Hitherto, we have discussed a single equipment with no redundancy (superfluous component). Now we consider a general case, where we have a single component with a single standby redundant component which has the same failure distribution λ as that of the original component. Here the redundant unit is not expected to fail unless it is put on the line (cold standby). As usual, we shall assume that when online equipment fails, the service time needed to bring back to working condition is negative exponentially distributed with parameter μ. However, if redundant unit is available then instantaneous replacement is possible with this set up. With usual notations we can write the Kolmogorov equations

$$p_0(t+\delta t) = p_0(t)[1-(\lambda t + 0(\delta t))] + p_1(t)[\mu \delta t + 0(\delta t)] \quad (3.6.1)$$

$$\Rightarrow \lim_{\delta t \to 0} \frac{p_0(t+\delta t) - p_0(t)}{\delta t} = -\lambda p_0(t) + \mu p_1(t) + \lim_{\delta t \to 0} \frac{0(\delta t)}{\delta t}$$

$$\Rightarrow p_0'(t) = -\lambda p_0(t) + \mu p_1(t). \quad (3.6.2)$$

Similarly,

$$p_1(t+\delta t) = p_1(t)[1-(\lambda+\mu)\delta t + 0(\delta t)] + p_0(t)[\lambda \delta t + 0(\delta t)]$$

$$\Rightarrow \lim_{\delta t \to 0} \frac{p_1(t+\delta t) - p_1(t)}{\delta t} = -(\lambda+\mu) p_1(t) + \lambda p_0(t) + \lim_{\delta t \to 0} \frac{0(\delta t)}{\delta t}$$

$$\Rightarrow p_1'(t) = -(\lambda+\mu) p_1(t) + \lambda p_0(t). \quad (3.6.3)$$

Finally,

$$p_2(t+\delta t) = p_1(t)[\lambda \delta t + 0(\delta t)] + p_2(t)$$

$$\Rightarrow \lim_{\delta t \to 0} \frac{p_2(t+\delta t) - p_2(t)}{\delta t} = \lambda p_1(t) + \lim_{\delta t \to 0} \frac{0(\delta t)}{\delta t}$$

$$\Rightarrow p_2'(t) = \lambda p_1(t). \quad (3.6.4)$$

From (3.6.2) we have,

$$L(p_0'(t)) = -\lambda L(p_0(t)) + \mu L(p_1(t)).$$

From (3.6.3),

$$L(p_1'(t)) = -(\lambda+\mu) L(p_1(t)) + \lambda L(p_0(t)).$$

From (3.6.4),

$$\begin{aligned} L(p_2'(t)) &= \lambda L(p_1(t)) \\ -1 + sL(p_0(t)) &= -\lambda L(p_0(t)) + \mu L(p_1(t)) \\ (s+\lambda) L(p_0(t)) - \mu L(p_1(t)) - 1 &= 0. \end{aligned} \quad (3.6.5)$$

Similarly from (3.6.3),

$$\begin{aligned} sL(p_1(t)) &= -(\lambda+\mu) L(p_1(t)) + \lambda L(p_0(t)) \\ (s+\lambda+\mu) L(p_1(t)) &= \lambda L(p_0(t)). \end{aligned} \quad (3.6.6)$$

Finally,

$$\begin{aligned} sL(p_2(t)) &= -\lambda L(p_2(t)) \\ (s+\lambda) L(p_0(t)) &= -\lambda L(p_1(t)) = 1. \end{aligned} \quad (3.6.7)$$

$$-\lambda L(p_0(t)) + (s+\lambda+\mu) L(p_1(t)) = 0 \quad (3.6.8)$$

$$-\lambda L(p_1(t)) + sL(p_2(t)) = 0. \quad (3.6.9)$$

$$L(p_0(t)) = \frac{\begin{vmatrix} 1 & -\mu & 0 \\ 0 & s+\lambda+\mu & 0 \\ 0 & -\lambda & s \end{vmatrix}}{\begin{vmatrix} s+\lambda & -\mu & 0 \\ -\lambda & s+\lambda+\mu & 0 \\ 0 & -\lambda & s \end{vmatrix}} = \frac{(s+\lambda+\mu)}{(s^2+s(2\lambda+\mu)+\lambda^2)}. \quad (3.6.10)$$

$$L(p_1(t)) = \frac{\begin{vmatrix} s+\lambda & 1 & 0 \\ -\lambda & 0 & 0 \\ 0 & 0 & s \end{vmatrix}}{\begin{vmatrix} s+\lambda & -\mu & 0 \\ -\lambda & s+\lambda+\mu & 0 \\ 0 & -\lambda & s \end{vmatrix}} = \frac{\lambda s}{s[s^2+s(2\lambda+\mu)+\lambda^2]}. \quad (3.6.11)$$

Similarly,

$$L(p_2(t)) = \frac{\lambda^2}{s[s^2+s(2\lambda+\mu)+\lambda^2]}. \quad (3.6.12)$$

We can write (3.6.10) as

$$\frac{s+\lambda+\mu}{(s-\xi_1)(s-\xi_2)} = \frac{A}{(s-\xi_1)} + \frac{B}{(s-\xi_2)},$$

where ξ_1 and ξ_2 are the roots of $s^2+s(2\lambda+\mu)+\lambda^2 = 0$

$$\Rightarrow \frac{\xi_1+\lambda+\mu}{(\xi_1-\xi_2)} = A, \quad \frac{\xi_2+\lambda+\mu}{(\xi_2-\xi_1)} = B.$$

$$\therefore p_0(t) = \frac{\xi_1+\lambda+\mu}{(\xi_1-\xi_2)}e^{\xi_1 t} - \frac{\xi_2+\lambda+\mu}{(\xi_1-\xi_2)}e^{\xi_2 t}. \quad (3.6.13)$$

Similarly, (3.6.11) can be written as

$$\begin{aligned} L(p_1(t)) &= \frac{\lambda}{s^2+s(2\lambda+\mu)+\lambda^2} = \frac{\lambda}{(s-\xi_1)(s-\xi_2)} \\ &= \frac{A}{(s-\xi_1)} + \frac{B}{(s-\xi_2)} \Rightarrow \frac{\lambda}{(\xi_1-\xi_2)} = A, \quad \frac{-\lambda}{(\xi_1-\xi_2)} = B. \end{aligned}$$

$$\therefore p_1(t) = \frac{\lambda}{(\xi_1-\xi_2)}e^{\xi_1 t} - \frac{\lambda}{(\xi_1-\xi_2)}e^{\xi_2 t}. \quad (3.6.14)$$

$$R(t) = p_0(t) + p_1(t) = \frac{1}{(\xi_1-\xi_2)}\left[(\xi_1+2\lambda+\mu)e^{\xi_1 t} - (\xi_2+2\lambda+\mu)e^{\xi_2 t}\right].$$

Now $\xi_1+\xi_2 = -(2\lambda+\mu)$ and $\xi_1\xi_2 = \lambda^2$, $\xi_1, \xi_2 < 0$.

$$\therefore R(t) = \frac{1}{(\xi_1-\xi_2)}\left[\xi_1 e^{\xi_2 t} - \xi_2 e^{\xi_1 t}\right]. \quad (3.6.15)$$

$$\begin{aligned} E(T) &= \int_0^\infty R(t)\,dt = \frac{1}{(\xi_1-\xi_2)}\left(\xi_1\int_0^\infty e^{\xi_2 t}\,dt - \xi_2\int_0^\infty e^{\xi_1 t}\,dt\right) \\ &= \frac{1}{(\xi_1-\xi_2)}\left(-\frac{\xi_1}{\xi_2}+\frac{\xi_2}{\xi_1}\right) = \frac{1}{(\xi_1-\xi_2)}\left(-\frac{\xi_1^2}{\xi_2^2}-\frac{\xi_2}{\xi_1}\right) \\ &= -\frac{(\xi_1+\xi_2)}{\xi_1\xi_2} = \frac{2\lambda+\mu}{\lambda^2} = \frac{2}{\lambda}+\frac{\mu}{\lambda^2} > 0. \end{aligned} \quad (3.6.16)$$

Also from (3.6.15) $\Rightarrow$

$$F(t) = 1 - \frac{1}{(\xi_1 - \xi_2)}\left[\xi_1 e^{\xi_2 t} - \xi_2 e^{\xi_t}\right] \tag{3.6.17}$$

$$f(t) = \frac{d}{dt}\left(1 - \frac{1}{(\xi_1 - \xi_2)}\left[\xi_1 e^{\xi_2 t} - \xi_2 e^{\xi_1 t}\right]\right).$$

Alternatively,

$$\begin{aligned} f(t) &= \frac{d}{dt}(1 - [p_0(t) + p_1(t)]) = \frac{d}{dt}p_2(t) \; (\because \; p_0(1) + p_1(t) + p_2(t) = 1) \\ &= p_2'(t) = \lambda p_1(t) = \lambda\left(\frac{\lambda}{(\xi_1 - \xi_2)}e^{\xi_1 t} - \frac{\lambda}{(\xi_1 - \xi_2)}e^{\xi_2 t}\right) \text{ [from (3.6.14)]} \\ &= \frac{\lambda^2}{(\xi_1 - \xi_2)}\left[e^{\xi_1 t} - e^{\xi_2 t}\right] \end{aligned} \tag{3.6.18}$$

3.7 Maintained System with Redundancy—Redundant Unit failing at a less Hazard Rate (Warm Standby)

In this case let us assume that λ_1 and λ_2 be the failure rates of the original component and standby ($\lambda_1 < \lambda_2$) and μ is the intensity of the common repair or servicing for bringing either the original or the redundant component in working condition. In this case we have the following Kolmogorov equations.

$$p_0(t + \delta t) = p_0(t)[1 - (\lambda_1 + \lambda_2)\delta t + 0(\delta t)] + p_1(t)[\mu\delta t + 0(\delta t)] \tag{3.7.1}$$

$$p_1(t + \delta t) = p_1(t)[1 - (\lambda_1 + \mu)\delta t + 0(\delta t)] + p_0(t)[(\lambda_1 + \lambda_2)\delta t + 0(\delta t)] \tag{3.7.2}$$

$$p_2(t + \delta t) = p_1(t)[\lambda_1\delta t + 0(\delta t)] + p_2(t). \tag{3.7.3}$$

►► NOTE

By putting $\lambda_1 = \lambda$ and $\lambda_2 = 0$ we go back to the model of (3.5) and putting $\mu = 0$ leads to example (3.10)

$$\Rightarrow \lim_{\delta t \to 0} \frac{p_0(t + \delta t) - p_0(t)}{\delta t} = -(\lambda_1 + \lambda_2)p_0(t) + \mu p_1(t) + \lim_{\delta t \to 0}\frac{0(\delta t)}{\delta t}$$

$$\Rightarrow p_0'(t) = -(\lambda_1 + \lambda_2)p_0(t) + \mu p_1(t). \tag{3.7.4}$$

Similarly, from (3.7.2),

$$\lim_{\delta t \to 0} \frac{p_1(t + \delta t) - p_1(t)}{\delta t} = -(\lambda_1 + \mu)p_1(t) + (\lambda_1 + \lambda_2)p_0(t) + \lim_{\delta t \to 0}\frac{0(\delta t)}{\delta t}$$

$$\Rightarrow p_1'(t) = -(\lambda_1 + \mu)p_1(t) + (\lambda_1 + \lambda_2)p_0(t). \tag{3.7.5}$$

From (3.7.3),

$$\lim_{\delta t \to 0} \frac{p_2(t + \delta t) - p_2(t)}{\delta t} = \lambda_1 p_1(t) + \lim_{\delta t \to 0}\frac{0(\delta t)}{\delta t}$$

$$\Rightarrow p_2'(t) = \lambda_1 p_1(t). \tag{3.7.6}$$

(3.7.4) $\Rightarrow$

$$-1 + sL(p_0(t)) = -(\lambda_1 + \lambda_2)L(p_0(t)) + \mu L(p_1(t))$$

$$(s + \lambda_1 + \lambda_2)L(p_0(t)) - \mu L(p_1(t)) + 0(L(p_2(t))) = 1. \tag{3.7.7}$$

(3.7.5) $\Rightarrow$

$$sL(p_1(t)) = -(\lambda_1 + \mu)L(p_1(t)) + (\lambda_1 + \lambda_2)L(p_0(t))$$

$$(\lambda_1 + \lambda_2)L(p_0(t)) - (s + \lambda_1 + \mu)L(p_1(t)) + 0(L(p_2(t))) = 0. \tag{3.7.8}$$

(3.7.6) $\Rightarrow$

$$sL(p_2(t)) = \lambda_1 L(p_1(t)). \tag{3.7.9}$$

On solving (3.7.7) and (3.7.8),

$$(s + \lambda_1 + \lambda_2)L(p_0(t)) - \mu L(p_1(t)) + 0(L(p_2(t))) = 1$$

$$(\lambda_1 + \lambda_2)L(p_0(t)) - (s + \lambda_1 + \mu)L(p_1(t)) + 0(L(p_2(t))) = 0$$

$$0(L(p_2(t))) - \lambda_1 L(p_1(t)) + sL(p_2(t)) = 0$$

$$\begin{aligned}
L(p_0(t)) &= \frac{\begin{vmatrix} 1 & -\mu & 0 \\ 0 & -(s+\lambda_1+\mu) & 0 \\ 0 & -\lambda_1 & s \end{vmatrix}}{\begin{vmatrix} s+\lambda_1+\lambda_2 & -\mu & 0 \\ \lambda_1+\lambda_2 & -(s+\lambda_1+\mu) & 0 \\ 0 & -\lambda_1 & s \end{vmatrix}} \\
&= \frac{-s(s+\lambda_1+\mu)}{s[-(s+\lambda_1+\lambda_2)(s+\lambda_1+\mu) + \mu(\lambda_1+\lambda_2)]} \\
&= \frac{(s+\lambda_1+\mu)}{(s+\lambda_1+\lambda_2)(s+\lambda_1+\mu) + \mu(\lambda_1+\lambda_2)} \\
&= \frac{(s+\lambda_1+\mu)}{s^2 + s(\lambda_1+\lambda_2) + s(\lambda_1+\mu) - \mu(\lambda_1+\lambda_2) + (\lambda_1+\mu)(\lambda_1+\lambda_2)} \\
&= \frac{(s+\lambda_1+\mu)}{s^2 + s(\lambda_1+2\lambda_2+\mu) + \lambda_1(\lambda_1+\lambda_2)} \\
&= \frac{s+\lambda_1+\mu}{(s-\xi_1)(s-\xi_2)} = \frac{A}{(s-\xi_1)} + \frac{B}{(s-\xi_2)}, \ (say),
\end{aligned}$$

where ξ_1 and ξ_2 are the roots of $s^2 + s(\lambda_1 + 2\lambda_2 + \mu) + \lambda_1(\lambda_1 + \lambda_2) = 0$.

$$A + B \lim_{s \to \xi_1} \frac{(s-\xi_1)}{(s-\xi_2)} \Rightarrow A = \frac{\xi_1 + \lambda_1 + \mu}{(\xi_1 - \xi_2)}.$$

Precisely, in the same way,

$$B = \frac{\xi_2 + \lambda_1 + \mu}{(\xi_2 - \xi_1)}.$$

$$p_0(t) = \frac{\xi_1 + \lambda_1 + \mu}{(\xi_1 - \xi_2)} e^{\xi_1 t} - \frac{\xi_2 + \lambda_1 + \mu}{(\xi_1 - \xi_2)} e^{\xi_2 t} \tag{3.7.10}$$

$$
\begin{aligned}
L(p_1(t)) &= \frac{\begin{vmatrix} s+\lambda_1+\mu & 1 & 0 \\ \lambda_1+\lambda_2 & 0 & 0 \\ 0 & 0 & s \end{vmatrix}}{\begin{vmatrix} s+\lambda_1+\lambda_2 & -\mu & 0 \\ \lambda_1+\lambda_2 & -(s+\lambda_1+\mu) & 0 \\ 0 & -\lambda_1 & s \end{vmatrix}} \\
&= \frac{-s(\lambda_1+\lambda_2)}{s[-(s+\lambda_1+\lambda_2)(s+\lambda_1+\mu)+\mu(\lambda_1+\lambda_2)]} \\
&= \frac{(\lambda_1+\lambda_2)}{(s+\lambda_1+\lambda_2)(s+\lambda_1+\mu)-\mu(\lambda_1+\lambda_2)} \\
&= \frac{(\lambda_1+\lambda_2)}{s^2+s(\lambda_1+\mu)+(\lambda_1+\lambda_2)(\lambda_1+\mu)-\mu(\lambda_1+\lambda_2)+s(\lambda_1+\lambda_2)} \\
&= \frac{(\lambda_1+\lambda_2)}{s^2+s(2\lambda_1+\lambda_2+\mu)+\lambda_1(\lambda_1+\lambda_2)} = \frac{(\lambda_1+\lambda_2)}{(s-\xi_1)(s-\xi_2)}. \\
L(p_1(t)) &= (\lambda_1+\lambda_2)\left(\frac{1}{(s-\xi_1)}-\frac{1}{(s-\xi_2)}\right)\frac{1}{(\xi_1-\xi_2)} \\
p_1(t) &= \frac{\lambda_1+\lambda_2}{\xi_1-\xi_2}\left(e^{\xi_1 t}-e^{\xi_2 t}\right). \qquad (3.7.11)
\end{aligned}
$$

From (3.7.10) and (3.7.11),

$$
\begin{aligned}
R(t) &= p_0(t)+p_1(t) \\
&= \frac{(\xi_1+\lambda_1+\mu)e^{\xi_1 t}-(\xi_2+\lambda_1+\mu)e^{\xi_2 t}+(\lambda_1+\lambda_2)(e^{\xi_1 t}-e^{\xi_2 t})}{\xi_1-\xi_2} \\
&= \frac{\xi_1 e^{\xi_2 t}-\xi_2 e^{\xi_1 t}}{\xi_1-\xi_2}. \qquad (3.7.12)
\end{aligned}
$$

ξ_1 and ξ_2 are the two roots of the quadratic equation

$$s^2+s(2\lambda_1+\lambda_2+\mu)+\lambda_1(\lambda_1+\lambda_2)=0$$

$$
\begin{aligned}
\xi_1 &= -\frac{(2\lambda_1+\lambda_2+\mu)+\sqrt{(2\lambda_1+\lambda_2+\mu)^2-4\lambda_1(\lambda_1+\lambda_2)}}{2} \\
&= \frac{(2\lambda_1+\lambda_2+\mu)+\sqrt{(\lambda_2^2+\mu^2+2\lambda_2\mu+4\lambda_1\mu}}{2} \\
\xi_2 &= -\frac{(2\lambda_1+\lambda_2+\mu)-\sqrt{(\lambda_2^2+\mu^2+2\lambda_2\mu+4\lambda_1\mu}}{2}
\end{aligned}
$$

$$\xi_1+\xi_2 = -\frac{2(2\lambda_1+\lambda_2+\mu)}{2} = -(2\lambda_1+\lambda_2+\mu)$$

$$\xi_1\xi_2 = \lambda_1(\lambda_1+\lambda_2).$$

Expected longevity of the system

$$
\begin{aligned}
E(T) = \int_0^\infty R(t)\,dt &= \frac{\xi_1}{\xi_1-\xi_2}\left(\frac{e^{\xi_2 t}}{\xi_2}\right)_0^\infty - \frac{\xi_2}{\xi_1-\xi_2}\left(\frac{e^{\xi_1 t}}{\xi_1}\right)_0^\infty \\
&= \frac{\xi_2}{\xi_1-\xi_2}\times\frac{1}{\xi_1}-\frac{\xi_1}{\xi_1-\xi_2}\frac{1}{\xi_2} = -\frac{\xi_1+\xi_2}{\xi_1\xi_2}.
\end{aligned}
$$

$$E(T) = -\frac{\xi_1 + \xi_2}{\xi_1 \xi_2} = \frac{(2\lambda_1 + \lambda_2 + \mu)}{\lambda_1(\lambda_1 + \lambda_2)}. \tag{3.7.13}$$

3.8 Steady-State Solution of the System under Instantaneous Repairability

A system or (components) are repaired as soon as they fail with intensity μ, then the long-term availability of the system is given by the *limiting distribution of the system.*

We have the Kolmogorov equations of the system considered in Section 3.7 subject to instantaneous repairment facility as

$$\begin{aligned} p_0'(t) &= -(\lambda_1 + \lambda_2)p_0(t) + \mu p_1(t) \\ p_1'(t) &= -(\lambda_1 + \mu)p_1(t) + (\lambda_1 + \lambda_2)p_0(t) + 2\mu p_2(t) \\ \text{and } p_2'(t) &= \lambda_1 p_1(t) - 2\mu p_2(t) \end{aligned}$$

If $\lim_{t\to\infty} p_n(t) = \pi_n,\ n = 0, 1, 2, \ldots$

$$\text{and} \quad p_0'(t) = p_1'(t) = p_2'(t) = 0 \text{ as } t \to \infty \tag{3.8.1}$$

then we have the equations corresponding to the steady state solution as

$$0 = -(\lambda_1 + \lambda_2)\pi_0 + \mu\pi_1$$

$$\Rightarrow (\lambda_1 + \lambda_2)\pi_0 = \mu\pi_1 \tag{3.8.2}$$

$$\Rightarrow \pi_0 = \frac{\mu\pi_1}{(\lambda_1 + \lambda_2)}. \tag{3.8.3}$$

Also in view of (3.8.1) and (3.8.2)

$$\lambda_1\pi_1 - 2\mu\pi_2 = 0 \Rightarrow \pi_2 = \frac{\lambda_1\pi_1}{2\mu}. \tag{3.8.4}$$

Since

$$\pi_0 + \pi_1 + \pi_2 = 1, \text{ identically} \tag{3.8.5}$$

putting (3.8.3) and (3.8.4) in (3.8.5) we get,

$$\frac{\mu\pi_1}{(\lambda_1 + \lambda_2)} + \pi_1 + \frac{\lambda_1\pi_1}{2\mu} = 1$$

$$\Rightarrow \pi_1\left(\frac{\mu}{\lambda_1 + \lambda_2} + 1 + \frac{\lambda_1}{2\mu}\right) = 1$$

$$\begin{aligned} \Rightarrow \pi_1 &= \frac{1}{\dfrac{\mu}{\lambda_1 + \lambda_2} + 1 + \dfrac{\lambda_1}{2\mu}} = \frac{(\lambda_1 + \lambda_2)2\mu}{2\mu^2 + 2\mu(\lambda_1 + \lambda_2) + \lambda_1(\lambda_1 + \lambda_2)} \\ &= \frac{2\mu(\lambda_1 + \lambda_2)}{2\mu^2 + (2\mu + \lambda_1)(\lambda_1 + \lambda_2)}. \end{aligned} \tag{3.8.6}$$

$$\begin{aligned} \pi_0 &= \frac{\mu}{\lambda_1 + \lambda_2} \times \frac{2\mu(\lambda_1 + \lambda_2)}{2\mu^2 + (2\mu + \lambda_1)(\lambda_1 + \lambda_2)} \\ &= \frac{2\mu^2}{2\mu^2 + (2\mu + \lambda_1)(\lambda_1 + \lambda_2)} \end{aligned} \tag{3.8.7}$$

and finally

$$\begin{aligned}\pi_2 &= \frac{\lambda_1}{2\mu} \times \frac{2\mu(\lambda_1+\lambda_2)}{2\mu^2+(2\mu+\lambda_1)(\lambda_1+\lambda_2)} \\ &= \frac{\lambda_1(\lambda_1+\lambda_2)}{2\mu^2+(2\mu+\lambda_1)(\lambda_1+\lambda_2)}. \end{aligned} \tag{3.8.8}$$

π_2 represents the long term expected proportion of time that the system will remain inoperative (known as *Down time*) due to long term unavailability.

3.9 Some Miscellaneous Problems in Reliability—Preventive Maintenance Policy (PMP), Age Replacement Policy (ARP)

Here we introduce the policy of completely replacing all the components of a machine (or a system or plant) periodically to get rid of catastrophic failure or breakdown of the system.

Let T be the period of age replacement. This means that every component is replaced on its natural failure or at the age T whichever is earlier. If $F_T(\cdot)$ be the failure cumulative distribution function under preventive maintenance policy (PMP/ARP), then

$$F_T(x) = \begin{cases} F(x) & \text{if } 0 \le x \le T \\ 1 & \text{if } x > T, \end{cases}$$

where $F(\cdot)$ is the natural failure cumulative distribution function of the component.

The waiting time Y between two consecutive natural failures is given by

$$Y = NT + Z, \tag{3.9.1}$$

where $N \equiv$ random number of replacement of the component by PMP/ARP; $N = 0, 1, 2, \ldots$ (a discrete random variable). $Z \equiv$ residual survival time following last PMP/ARP (an absolutely continuous random variable).

$$E(Y) = E(N)T + E(Z). \tag{3.9.2}$$

Now

$$E(N) = \sum_{n=1}^{\infty} P(N = n)n. \tag{3.9.3}$$

Also

$$\begin{aligned} P(N = n) &= P(N \ge n) - P(N \ge n+1), \text{ since } N \text{ is discrete} \\ &= [1 - F(T)]^n - [1 - F(T)]^{n+1}. \end{aligned} \tag{3.9.4}$$

Putting (3.9.4) in (3.9.3), we have

$$\begin{aligned} E(N) &= \sum_{n=1}^{\infty} n\left[\{1 - F(T)\}^n - \{1 - F(T)\}^{n+1}\right] \\ &= \left[\{1 - F(T)\} - \{1 - F(T)\}^2\right] + 2\left[\{1 - F(T)\}^2 - \{1 - F(T)\}^3\right] + \ldots \\ &= \{1 - F(T)\} + \{1 - F(T)\}^2 + \ldots \\ &= \{1 - F(T)\}\left[1 + \{1 - F(T)\} + \{1 - F(T)\}^2 + \ldots\right] \\ &= \{1 - F(T)\}\left[1 - \{1 - F(T)\}\right]^{-1} \\ &= \frac{1 - F(T)}{F(T)}. \end{aligned} \tag{3.9.5}$$

Again

$$F_T(z) = P[Z \le z] = \frac{F(z)}{F(T)}, \text{ under PMP/ARP.} \tag{3.9.6}$$

$$R_T(z) = 1 - F_T(z) = 1 - \frac{F(z)}{F(T)}$$

$$E(z) = \int_0^T \left(1 - \frac{F(z)}{F(T)}\right) dz = \frac{TF(T) - \int_0^T F(z)\,dz}{F(T)} \tag{3.9.7}$$

Putting (3.9.5) and (3.9.7) in (3.9.2),

$$\begin{aligned} E(Y) &= \frac{TF(T) - \int_0^T F(z)\,dz}{F(T)} + \frac{T\{1 - F(T)\}}{F(T)} \\ &= \frac{T - \int_0^T F(z)\,dz}{F(T)} = \frac{\int_0^T (1 - F(z))\,dz}{F(T)} \end{aligned} \tag{3.9.8}$$

represents the expected time between true time of failures under the PMP/ARP.

Example 11 *Show that the true time to failure under* PMP/ARP *at negative exponential failure rate is the true time without* PMP/ARL, *and* PMP *has no advantage if negative exponential failure rate occurs.*

Solution: We have

$$f(x) = \lambda e^{-\lambda x};\ \lambda > 0,\ 0 < x < \infty$$

$$F(x) = 1 - e^{-\lambda x}$$

$$E(T) = \frac{\int_0^T 1 - [1 - e^{\lambda z}]\,dz}{1 - e^{-\lambda T}}$$

$$E(T) = \frac{\int_0^T e^{-\lambda z}\,dz}{1 - e^{-\lambda T}} = \frac{e^{-\lambda z}/(-\lambda)\Big|_0^T}{1 - e^{-\lambda T}} = \frac{1 - e^{-\lambda T}}{\lambda(1 - e^{-\lambda T})} = \frac{1}{\lambda},$$

whereas the expected time to natural failure for negative exponential distribution unrestricted is

$$\int_0^\infty x\lambda e^{-\lambda x}\,dx = \frac{1}{\lambda}.$$

3.10 An Availability Problem (steady-state solution)

Suppose, we have a multicomponent system of M components and m be the minimum number of components which should be in working order (i.e., the system will work even if $M - m$ components are out of order). Suppose, the components have negative exponential life distribution with failure rate λ. Further at most s number of components can be repaired simultaneously and the mean repair time is $1/\mu$, where μ is a parameter of a negative exponential distribution.

Let $N(t)$ be the number of components inoperative at time t and

$$p_n(t) = P[N(t) = n]$$

$p_n(t)$ represents the availability of $(M - n)$ components at time t.

If $n \geq M - m$, the system is not available.

If $n < N - m$, the system is available.

The *Stochastic process* $N(t)$ is a birth and death process with birth and death parameters λ_n an μ_n respectively depending on the initial size of the population n.

$$\lambda_n = (M - n)\lambda.$$

$$\mu_n = \begin{cases} n\mu & \text{if } 1 \leq n \leq s-1 \\ s\mu & \text{if } n \geq s. \end{cases}$$

The Kolmogorov equations of the birth and death process are as follows:

$$p_0'(t) = -M\lambda p_0(t) + \mu p_1(t) \tag{3.10.1}$$

$$p_n'(t) = -[(M-n)\lambda + n\mu]p_n(t) + (M-n+1)_\lambda p_{n-1}(t) + (n+1)\mu p_{n+1}(t) \tag{3.10.2}$$

for $1 \leq n \leq s - 1$.

Also

$$p_n'(t) = -[(M-n)\lambda + s\mu]p_n(t) + (M-n+1)\lambda p_{n-1}(t) + s\mu p_{n+1}(t) \tag{3.10.3}$$

for $s \leq n \leq M$ and finally

$$p_M'(t) = -s\mu p_M(t) + \lambda p_{M-1}(t). \tag{3.10.4}$$

The steady-state solution of (3.10.2) is given by assuming $p_n'(t) = 0$, $n = 0, 1, \ldots$

$$\lim_{n\to\infty} p_n(t) = \pi_n$$

$$(3.10.2) \Rightarrow M\lambda\pi_0 = \mu\pi_1 \Rightarrow \pi_1 = \frac{M\lambda}{\mu}\pi_0 \tag{3.10.5}$$

$$(3.10.2) \Rightarrow [(M-n)\lambda + n\mu]\pi_n = (M-n+1)\lambda\pi_{n-1} + (n+1)\mu\pi_{n+1} \tag{3.10.6}$$

for $1 \leq n \leq s$.

$$(3.10.3) \Rightarrow [(M-n)\lambda + s\mu]\pi_n = (M-n+1)\lambda\pi_{n-1} + s\mu\pi_{n+1} \tag{3.10.7}$$

for $s \leq n \leq M$.

$$(3.10.4) \Rightarrow s\mu\pi_M = \lambda\pi_{M-1}. \tag{3.10.8}$$

Solving (3.10.5), (3.10.6), (3.10.7) and (3.10.8)

$$\Rightarrow \quad \pi_n = \binom{M}{n}\theta^n\pi_0, \text{ where } \theta = \lambda/\mu,\ 0 \leq n \leq s$$

$$\Rightarrow \quad \pi_n = \binom{M}{n}\frac{n!}{s!s^{n-s}}\theta^n\pi_0,\ s \leq n \leq M - m + 1. \tag{3.10.9}$$

Further

$$\sum_{n=0}^{M}\pi_n = 1. \tag{3.10.10}$$

In particular when $s = 1$

$$\Rightarrow \pi_n = \frac{M!}{(M-n)!}\theta^n \pi_0 \tag{3.10.11}$$

$$\pi_0 = \left[1 + \frac{M!}{(M-1)!}\theta + \frac{M!}{(M-2)!}\theta^2 + \cdots + \frac{M!}{(m-1)!}\theta^{M-m+1}\right]^{-1} \tag{3.10.12}$$

for $1 \leq n \leq M - m + 1$.

Chapter 4

Maintenance

4.1 Certain Basic Results

Increasing Failure Rate (IFR): We call the system (or component) to have IFR if the hazard function at any time denoted by $h(t)$ is an increasing function of t.

Decreasing Failure Rate (DFR): Similarly, a system is said to have DFR if the hazard function $h(t)$ at any time is a decreasing function of t.

Increasing Failure Rate on the Average (IFRA):

$$\frac{\int_0^t h(\tau)\,d\tau}{t}$$

is an increasing function of t

$$\Rightarrow \int_0^t h(\tau)\,d\tau$$

is also an increasing function of t, which in other words can be stated as equal to $-\log R(t)/t$ to be an increasing function of t for $t \geq 0$, where

$$R(t) = \exp\left[-\int_0^t h(\tau)\,d\tau\right] \Rightarrow [-\log R(t)/t]$$

to be an increasing function of t is the definition of IFRA, where $R(t)$ be the Reliability at any time t. Thus we have,

$$\text{IFR} \Rightarrow \text{IFRA}.$$

Similarly, DFR $\Rightarrow$ DFRA, where DFRA (decreasing failure rate on the average) is defined as $-\log R(t)/t$ to be a decreasing function of t.

New better than used (NBU): A system (or component) is said to be NBU if the probability of surviving up to $(x+y)$ cannot exceed the probability of surviving up to x and then (considering the system to be a renewed one) probability of surviving up to y. Symbolically, we have,

$$R(x+y) \leq R(x)R(y).$$

We can see that

$$\begin{aligned} \text{IFRA} &\Rightarrow \text{NBU} \\ \text{and DFRA} &\Rightarrow \text{NWU} \quad \text{(new worse than used).} \end{aligned}$$

Proof.

$$\begin{aligned} e^{-\int_0^{x+y} h(\tau)\,d\tau} &= e^{-\int_0^{x} h(\tau)\,d\tau} + e^{-\int_x^{x+y} h(\tau)\,d\tau} \\ &= e^{-\left[\int_0^{x} h(\tau)\,d\tau + \int_x^{x+y} h(\tau)\,d\tau\right]} \\ &= e^{-\int_0^{x} h(\tau)\,d\tau} \cdot e^{-\int_x^{x+y} h(\tau)\,d\tau} \end{aligned}$$

Again,

$$e^{-\int_0^{x} h(\tau)\,d\tau} \geq e^{-\int_x^{x+y} h(\tau)\,d\tau}, \ \forall\, x \geq 0,$$

if the system is IFRA

$$\therefore\ R(x+y) \leq R(x)R(y).$$

Thus,

$$\text{IFRA} \Rightarrow \text{NBU}.$$

Similarly,

$$\text{DFRA} \Rightarrow \text{NWU (New worse than used)}$$

New better than used in expectation (NBUE): We define NBUE as the residual mean longevity of a system beyond a period x given that the same has survived up to x (i.e., an old system's age since it has survived up to x) is less than the mean longevity of a system (i.e., new). Denoting by $R(x)$ the reliability at the age x, NBUE implies that,

$$\int_x^{\infty} R(\tau)\,d\tau / R(x) \leq \int_0^{\infty} R(\tau)\,d\tau = \mu,$$

the mean longevity of the system.

Again, it may be seen that NBU $\Rightarrow$ NBUE.

Since $R(x+y) \leq R(x) \cdot R(y)$ for NBU, it follows that

$$\int_0^{\infty} R(x+y)\,dy \leq R(x) \int_0^{\infty} R(y)\,dy.$$

Putting $x + y = \tau$, we have

$$\frac{\int_x^{\infty} R(\tau)\,d\tau}{R(x)} = \int_0^{\infty} R(y)\,dy = \mu,$$

the mean longevity of the system.

Thus

$$\text{NBU} \Rightarrow \text{NBUE}.$$

Combining all the results, we have

$$\text{IFR} \Rightarrow \text{IFRA} \Rightarrow \text{NBU} \Rightarrow \text{NBUE}.$$

Similarly, we can show that

$$\text{NBUE} \Rightarrow \text{NBU} \Rightarrow \text{IFRA} \Rightarrow \text{IFR}.$$

Therefore,

$$\text{IFR} \Leftrightarrow \text{IFRA} \Leftrightarrow \text{NBU} \Leftrightarrow \text{NBUE}.$$

Example 1 *Show that* NBUE $\Rightarrow$ NBU $\Rightarrow$ IFRA $\Rightarrow$ IFR.

4.2 Renewal Theory

Let $S_n = X_1 + X_2 + \cdots + X_n$ represent the waiting time for the nth renewal (or replacement or failure) and $X_i (i = 1, 2, \ldots, n)$ represent the waiting time from the $(i-1)$th renewal to the ith renewal. X_i's are independent random variables (independent identically distributed random variables).

Now $P[S_n \leq t] = P[N_t \geq n] = F_n(t)$, where N_t represents the number of renewals in $(0, t]$ and $F_n(t)$ is the cumulative distribution of S_n.

Then

$$\begin{aligned} P[N_t = n] &= P[N_t \geq n] - P[N_t \geq n+1] \\ &= F_n(t) - F_{n+1}(t). \end{aligned}$$

$$\begin{aligned} E(N_t) &= \sum_{n=0}^{\infty} nP[N_t = n] = \sum_{n=0}^{\infty} n[F_n(t) - F_{n+1}(t)] \\ &= 1(F_1(t) - F_2(t)) + 2(F_2(t) - F_3(t)) + \ldots \\ &= F_1(t) + F_2(t) + \cdots = \sum_{n=1}^{\infty} F_n(t). \end{aligned}$$

We define $E(N_t) = U(t)$ = renewal function or the average number of renewals in $(0, t]$

$$\frac{dE(N_t)}{dt} = \frac{dU(t)}{dt} = \frac{d}{dt} \sum_{n=1}^{\infty} F_n(t) = \sum_{n=1}^{\infty} \frac{d}{dt} F_n(t) = \sum_{n=1}^{\infty} f_n(t) = u(t),$$

where $f_n(t)$ is the interval density of the nth arrival and $u(t)$ is the renewal density which is the derivative of $U(t)$, the renewal function (i.e., the average number of renewals in $(0, t)$. Thus $u(t)\, dt$ represents the probability that a renewal (of any order) will take place between t to $t + dt$, whereas the interval density $f_n(t)\, dt$ represents the probability of the time interval lying between t to $t + dt$ following the arrival time of the $(n-1)$th renewal. The Laplace's transformation of $u(t)$ is given by

$$\begin{aligned} L(u(t)) &= \int_0^{\infty} e^{-st} u(t)\, dt = \int_0^{\infty} e^{-st} \sum_{n=1}^{\infty} f_n(t)\, dt \\ &= \sum_{n=1}^{\infty} \int_0^{\infty} e^{-st} f_n(t)\, dt = \sum_{n=1}^{\infty} L(f_n(t)). \end{aligned}$$

Let the distribution of the waiting time X_1 be $f_1(t)$ and the common distribution of X_i $(i = 2, 3, 4, \ldots)$ be $f(t)$.

Then

$$f_n(t) = f_1(t) * [f(t)]^{*(n-1)},$$

where $*(n-1)$ stands for $(n-1)$ fold convolution.

Therefore,

$$L(f_n(t)) = L[f_1(t)][L(f(t)]^{(n-1)}.$$

This can be proved as follows.

Let us consider the convolution of two functions $f_1(t)$ and $f_2(t)$ denoted by

$$\begin{aligned} f_1^* f_2(t) &= \int_0^t f_1(\tau) f_2(t-\tau)\, d\tau. \\ \therefore\ L(f_1^* f_2(t)) &= \int_0^\infty e^{-st} \left\{ \int_0^t [f_1(\tau) f_2(t-\tau)\, d\tau] \right\} dt \\ &= \int_0^\infty \int_0^t e^{-st} f_1(\tau) f_2(t-\tau)\, d\tau dt. \end{aligned}$$

Changing the order of integration,

$$\begin{aligned} \int_0^\infty e^{-s\tau} f_1(\tau) \int_\tau^\infty e^{-s(t-\tau)} f_2(t-\tau)\, dt d\tau &= \int_0^\infty e^{-s\tau} f_1(\tau)\, d\tau \cdot \int_\tau^\infty e^{-s(t-\tau)} f_2(t-\tau)\, d\tau \\ &= \int_0^\infty e^{-s\tau} f_1(\tau)\, d\tau \cdot \int_0^\infty e^{-su} f_2(u)\, du \\ &= L(f_1(t)) L(f_2(t)). \end{aligned}$$

Similarly, for n functions,

$$L(f_1^* f_2^* f_3^* \ldots {}^* f_n(t)) = L(f_1(t)) L(f_2(t)) \ldots L(f_n(t)).$$

Also

$$\begin{aligned} L(u(t)) &= \sum_{n=1}^\infty L(f_n(t)) = \sum_{n=1}^\infty L(f_1(t))[L(f(t)]^{n-1} \\ &= L(f_1(t)) \sum_{n=1}^\infty [L(f(t))]^{n-1} = \frac{L(f_1(t))}{1 - L(f(t))}. \end{aligned} \tag{4.2.1}$$

Next,

$$L(U(t)) = \int_0^\infty e^{-st} U(t)\, dt$$

and

$$\begin{aligned} L(u(t)) &= \int_0^\infty e^{-st} u(t)\, dt = e^{-st} U(t) \Big|_0^\infty + s \int_0^\infty e^{-st} U(t)\, dt \\ &= sL(U(t))\ (\because\ U(0) = 0). \end{aligned} \tag{4.2.2}$$

$$\begin{aligned} sL(U(t)) = \frac{L(f_1(t))}{1 - L(f(t))} \quad &\Rightarrow \quad sL(U(t)) = sL(U(t)) L(f(t)) = L(f_1(t)) \\ &\Rightarrow \quad L(U(t)) - L(U(t)) L(f(t)) = \tfrac{1}{s} L(f_1(t)). \end{aligned} \tag{4.2.3}$$

Again,

$$L(f_1(t)) = \int_0^\infty e^{-st} f_1(t)\, dt = e^{-st} F_1(t) \Big]_0^\infty + s \int_0^\infty e^{-st} F_1(t)\, dt,$$

where $F_1(t)$ is the cumulative distribution function corresponding to the density function $f_1(t)$.

$$\Rightarrow \quad L(f_1(t)) = sL(F_1(t)) \tag{4.2.4}$$

$$\text{or,} \quad \tfrac{1}{s} L(f_1(t)) = L(F_1(t)). \tag{4.2.5}$$

Putting (4.2.4) in (4.2.3) we have,

$$\Rightarrow L(U(t)) = L(U(t))L(f(t)) + L(F_1(t)).$$

Taking inverse Laplace's transformations on both sides, we have

$$\begin{aligned} U(t) &= F_1(t) + \int_0^t U(t-\tau)f(\tau)\,d\tau \\ &= F_1(t) + U(t)^* f(\tau), \end{aligned} \tag{4.2.6}$$

which is the forward renewal equation.

Differentiating both sides with respect to t, we have

$$u(t) = f_1(t) + \int_0^t u(t-\tau)f(\tau)\,d\tau. \tag{4.2.7}$$

(4.2.6) and (4.2.7) are both forward renewal equations; one concerning $U(t)$ and the other $u(t)$.

4.2.1 Alternative treatment for forward integral equations

Let τ be the time for the first renewal and N_t represents the number of renewals in $(0, t]$.

If (*i*) $t < \tau$, then $N_t = 0$, clearly

(*ii*) $t \geq \tau$, then

$$N_t = 1 + N_{t-\tau}. \tag{4.2.8}$$

Denote the probability generating function (p.g.f.) $E(s^{N_t})$ for any real integer variable N_t with $|\, s \,| < 1$ as $\Pi_t(s)$.

Thus

$$s^{N_t} = \begin{cases} 1, & \text{if } t < \tau \\ s^{1+N_{t-\tau}}, & \text{if } t \geq \tau. \end{cases}$$

$$\begin{aligned} \Pi_t(s) = E(S^{N_t}) &= E(E(S^{N_t} \mid \tau) = \int_0^\infty E(S^{N_t} \mid \tau)f(\tau)\,d\tau \\ &= \int_0^t E(S^{N_t} \mid \tau)f(\tau)\,d\tau + \int_t^\infty E(S^{N_t} \mid \tau)f(\tau)\,d\tau \\ &= \int_0^t E(S^{N_{t-\tau}+1})f(\tau)\,d\tau + \int_t^\infty f(\tau)\,d\tau. \\ \pi_t(s) &= s\int_0^t \pi_{t-\tau}(s)f(\tau)\,d\tau + \int_t^\infty f(\tau)\,d\tau, \end{aligned} \tag{4.2.9}$$

which is a recurrence relation between $\pi_t(s)$ and $\pi_{t-\tau}(s)$.

Differentiating both sides with respect to s and putting $s = 1$.

$$\left.\frac{d\Pi_t(s)}{ds}\right|_{s=1} = \int_0^t \Pi_{t-\tau}(s)|_{s=1} f(\tau)\,d\tau \int_0^t \left.\frac{d\pi_{t-\tau}(s)}{ds}\right|_{s=1} f(\tau(\tau\tau))$$

$$\Rightarrow U(t) = F(t) + \int_0^t U(t-\tau)f(\tau)\,d\tau. \tag{4.2.10}$$

This equation is a special case of (4.2.6) when $F_1 = F$ and $f_1 = f$. In fact, (4.2.6) refers to a general recurrent process, whereas (4.2.10) corresponds to a simple renewal process. Differentiating (4.2.10) with respect to

$$\Rightarrow u(t) = f(t) + \int_0^t u(t-\tau)f(\tau)\,d\tau, \tag{4.2.11}$$

which is a special case of (4.2.7).

4.3 Basic Renewal Theorems

Theorem 1 (Asymptotic Renewal Theorem) *If N_t represents the number of renewals up to time t, S_{N_t} represents the waiting time for N_t, the number of renewals, then*

$$P\left[\lim_{t\to\infty} \frac{S_{N_t}}{N_t}\right] = \mu,$$

where μ represents the average time between renewals.

Proof. We have

$$S_{N_t} \le t \le S_{N_t+1} \Rightarrow \frac{S_{N_t}}{N_t} \le \frac{t}{N_t} \le \frac{S_{N_t+1}}{N_{t+1}}\frac{N_{t+1}}{N_t}. \tag{4.3.1}$$

Given $N_t = n$, *say,* we have

$$\frac{S_{N_t}}{N_t} = \frac{X_1 + X_2 + \cdots + X_n}{n},$$

where $X_i s (i = 1, 2, \ldots, n)$ are independent identically distributed random variables with expectation existing and equal to μ.

Therefore, by strong law of large numbers, we have

$$\frac{S_{N_t}}{N_t} \xrightarrow{a.s.} \mu, \text{ i.e., } P\left[\lim_{t\to\infty} \frac{S_{N_t}}{N_t}\right] = \mu.$$

From (4.3.1) this implies

$$\frac{t}{N_t} \xrightarrow{a.s.} \mu, \text{ i.e., } P\left[\lim_{t\to\infty} \frac{N_t}{\mu}\right] = \frac{1}{\mu}. \tag{4.3.2}$$

This proves the result.

▶▶ NOTE

As strong consistency implies weak consistency, therefore,

$$\lim\left[P\left(\frac{N_t}{t}\right)\right] = \frac{1}{\mu}.$$

Theorem 2 (Elementary Renewal Theorem) *Given N_t/t converges almost surely (a.s.) to $1/\mu$ then*

$$\lim \frac{E(N_t)}{t} = \frac{1}{\mu} \text{ as } t \to \infty.$$

The result follows obviously since if N_t/t converges to a limit a.s. then it also converges to its expectation.

Definition 1 *A random variable X is said to be lattice if $\exists$ (there exists) one h, $\ni$ (such that)*

$$P[X = nh,\ n = 0, 1, 2, \ldots] = 1.$$

Theorem 3 (Blackwell's Theorem) *If F is a non-lattice distribution with mean μ, then*

$$\lim[U(t+h) - U(t)] = \frac{h}{\mu},$$

where $U(t)$ is the renewal function (average number of renewals up to t).

Proof. We have by elementary renewal theorem,

$$\lim_{t\to\infty} \frac{E(N_{t+h})}{t+h} \to \frac{1}{\mu} \text{ as } t \to \infty, \text{ i.e., } \lim_{t\to\infty} \frac{U_{t+h}}{t+h} = \frac{1}{\mu} \text{ as } t \to \infty.$$

Similarly,

$$\lim_{t\to\infty} \frac{U_t}{t} = \frac{1}{\mu} \text{ as } t \to \infty$$

$$\Rightarrow \quad \lim_{t\to\infty} [U_{t+h} - U_t] = \frac{t+h}{\mu} - \frac{t}{\mu} = \frac{h}{\mu} \text{ as } t \to \infty. \tag{4.3.3}$$

Corollay 1 *$u(t)$ = renewal function $= \frac{d}{dt}U(t) = \frac{1}{\mu}$ as $t \to \infty$ is obtained by proceeding to the limit $h \to 0$ in the Blackwell's theorem.*

Theorem 4 (Key Renewal Theorem) *If F is a non-lattice distribution with mean μ and $Q(t)$ is Riemann integrable (R-integrable), then*

$$\lim_{t\to\infty} \int_0^t Q(t-x)\, dU(x) = \frac{1}{\mu} \int_0^\infty Q(t)\, dt. \tag{4.3.4}$$

Proof. We have,

$$\int_0^t Q(t-x)\, dU(x) = \int_0^t Q(t-x) u(x)\, dx$$

since $Q(t)$ is R-integrable.

Putting

$$x = t - t' \Rightarrow dx = dt \Rightarrow . \int_0^t Q(t') u(t-t')\, dt'.$$

Again, as $t \to \infty$, $u(t-t')\, dt' \to 1/\mu$ by corollary to Elementary Renewal Theorem. Hence,

$$\int_0^t Q(t-x)\, dU(x) = \frac{1}{\mu} \int_0^\infty Q(t')\, dt' = \frac{1}{\mu} \int_0^\infty Q(t)\, dt \textbf{ (Proved)}.$$

4.4 Some Applications of the Results

4.4.1 Residual survival time for NBU and NWU

Given that a system is available at t, the probability that the system will survive a residual period

$$\gamma(t) = S_{N(t)+1} - t \text{ (a random variable) } \geq u, \ (say)$$

is denoted by $R_{\gamma(t)}(u)$. Similarly, the cumulative distribution function of $\gamma(t)$,

$$\text{i.e., } P[\gamma(t) \leq u] = F_{\gamma(t)}(u) = 1 - R_{\gamma(t)}(u).$$

We have the forward integral equation for residual survival time, given by

$$R_{\gamma(t)}(u) = R(t+u) + \int_0^t R(t+u-x)u(x)\,dx$$

subject to

$$R_{\gamma(t)}(0) = R(t) + \int_0^t R(t-x)u(x)\,dx.$$

$$\begin{aligned} P[\gamma(t) > u] = R_{\gamma(t)}(u) &= R(t+u) + \int_0^t R(t+u-x)u(x)\,dx \\ &\leq R(t)R(u) + \int_0^t R(t-x)R(u)u(x)\,dx, \text{ under NBU} \\ &= R(t)R(u) + R(u)\int_0^t R(t-x)u(x)\,dx \\ &= R(u)\left[R(t) + \int_0^t R(t-x)u(x)\,dx\right] \\ &= R(u)P[\gamma(t) > 0] = R(u). \end{aligned}$$

Thus, for NBU,

$$R_{\gamma(t)}(u) \leq R(u) \Rightarrow F_{\gamma(t)}(u) \geq F(u). \tag{4.4.1}$$

Similarly, for NWU,

$$R_{\gamma(t)}(u) \geq R(u) \Rightarrow F_{\gamma(t)}(u) \leq F(u). \tag{4.4.2}$$

4.4.2 A result on the average number of renewals under NBU/NWU

If F is NBU, then

$$U(t+h) - U(t) \geq U(h),$$

the inequality is reversed if F is NWU.
Proof. We have,

$$U(h) = \sum_{n=0}^{\infty} nP(N(h) = n) = \sum_{n=0}^{\infty} n\left[F^{(n)}(h) - F^{(n+1)}(h)\right] = \sum_{k=0}^{\infty} F^{(k+1)}(h),$$

where $F^{(n)}(h) = P[S_n \leq h] = P[N(h) \geq n]$.

Also

$$\begin{aligned} U(t+h) - U(t) &= \text{Average \# of renewals (replacements) between } t \text{ to } (t+h) \\ &= \sum_{k=0}^{\infty} F_{\gamma(t)}(u) F_{\gamma(t)}^{(k)}(t+u+h-(t+u)) \\ &= \sum_{k=0}^{\infty} F_{\gamma(t)}^{(k+1)}(h) \geq \sum_{k=0}^{\infty} F^{(k+1)}(h) \text{ [from (4.4.1)]} \end{aligned}$$

where $F_{\gamma(t)}^{(k)}(u) = P$ [at least k number of renewals/replacements up to time u and $F_{\gamma(t)}(u) = p$ [at least one replacement is up to u].

Also F

$$U(h) = \sum_{k=0}^{\infty} F^{(k+1)}(h).$$

Thus

$$U(t+h) - U(t) \geq U(h) \text{ for NBU.} \tag{4.4.3}$$

Precisely in a similar way,

$$U(t+h) - U(t) \leq U(h) \text{ for NWU.} \tag{4.4.4}$$

4.5 Stationary Renewal Process

Definition 2 *A process denoted by* $\{\bar{X}_k\}_{k=1}^{\infty}$ *is called a* Stationary renewal process, *if all random variables* $\hat{X}_1$, $\hat{X}_2$, ..., *are all independent with* $\hat{X}_2$, $\hat{X}_3$, ... *having a common cumulate distribution function* F *and mean* μ, *while* $\hat{X}_1$ *is distributed according to another cumulative distribution function* $\hat{F}(t)$ *which is related to* $F(t)$ *given by*

$$\hat{F}(t) = \frac{1}{\mu} \int_0^t R(x)\, dx,$$

where $R(x) = 1 - F(x)$. *Then* $\hat{N}(t)$ *for* $t \geq 0$ *is called a* Stationary Renewal Counting Process *corresponding to* $\hat{F}(t)$ *given by*

$$E(\hat{N}(t)) = \frac{t}{\mu}.$$

Differentiating $\hat{F}(t)$, with respect to t, we have,

$$\hat{f}(t) = \frac{1}{\mu} R(t) \Rightarrow \int_0^{\infty} f(t)\, dt = \frac{1}{\mu} \int_0^{\infty} R(t)\, dt = 1,$$

which shows that $\hat{f}(t)$ is a proper probability density.

4.5.1 Wald's equation

$$E\left[\sum_{i=1}^{N} X_i\right] = E(N)E(X_i)$$

provided X_i's are independent identically distributed ($i = 1, 2, \ldots, N$) and the even $N = n$ is independent of $X_{n+1}, X_{n+2}, \ldots$ for every $n = 1, 2, \ldots$.

Proof. Let us define random variables

$$Z_i = \begin{cases} 1, & \text{if } i \le N \\ 0, & \text{if } I > N \end{cases}$$

$$\sum_{i=1}^{N} X_i = \sum_{i=1}^{\infty} X_i Z_i \Rightarrow E\left(\sum_{i=1}^{N} X_i\right) = E\left(\sum_{i=1}^{\infty} X_i Z_i\right).$$

Now, since X_is are determined for $i \le N$ whereas Z_is are also determined by $i < N$, therefore, Z_is are independent of X_i. In view of the same

$$E\left(\sum_{i=1}^{\infty} X_i Z_i\right) = E(X) E\left(\sum_{i=1}^{\infty} Z_i\right),$$

where X_is being independent identically distributed is assumed to have been distributed as X.

$$\therefore\ E\left(\sum_{i=1}^{N} X_i\right) = E(X) \sum_{i=1}^{\infty} P(N \ge i) = E(X) E(N).$$

Example 2 $f_{X_i}(t) = \frac{1}{\mu} e^{-\frac{1}{\mu}t}$, $t = 1, 2, \ldots$, *then* $X_1, X_2, \ldots$ *conform to a stationary renewal process, i.e.,*

$$f_{x_i}(t) = \frac{R_i(x)}{\mu} \quad (i = 1, 2, \ldots)$$

Theorem 5 *F of a renewal process having mean* $\mu < \infty \Rightarrow U(t) \ge \frac{t}{\mu} - 1$.

Proof. We have the residual survival time

$$\gamma(t) = X_1 + X_2 + \cdots + X_{N(t)+1} - t,$$

where X_i is the waiting time between $(i-1)$th and ith replacement ($i = 1, 2, \ldots$)

$$\Rightarrow E(\gamma(t)) = E\left[\sum_{i=1}^{N(t)+1} X_i - t\right] = E(N(t)+1)E(X_i) - t$$

by Wald's identity provided X_is (i.i.d.r.v.s.) and $N(t)$ are independently distributed.

$$\therefore\ E(\gamma(t)) = (U(t)+1)\mu - t \ge 0 \Rightarrow U(t) \ge \frac{t}{\mu} - 1.$$

Theorem 6 *If F is* NBUE, *then* $\hat{F}(t) \ge F(t)$.

Proof. If F is NBUE, then

$$\frac{\int_t^{\infty} R(x)\, dx}{R(t)} \le \int_0^{\infty} R(x)\, dx = \mu,$$

where $R(x) = 1 - F(x)$ and μ is the mean survival period.

$$\Rightarrow \int_t^\infty R(x)\,dx \leq \mu R(t)$$

$$\Rightarrow \mu - \int_t^\infty R(x)\,dx \geq \mu - \mu R(t) = \mu(1 - R(t)) = \mu F(t).$$

$$\therefore \int_0^t R(x)\,dx \geq \mu F(t) \Rightarrow \frac{\int_0^t R(x)\,dx}{\mu} = \text{c.d.f. of } \bar{X}_1 = \hat{F}(t)$$

as per the notation of 4.5.

$$\Rightarrow \hat{F}(t) \geq F(t) \text{ under NBUE.} \tag{4.5.1}$$

Similarly,

$$\text{the inequality is reversed for NWUE.} \tag{4.5.2}$$

Theorem 7 *$\hat{U}(t) \geq U(t)$, where $\hat{U}(t)$ and $U(t)$ represent the average number of renewals in $(0, t]$ under $\hat{F}$ and F respectively.*

Now We have,

$$\hat{U}(t) = \sum_{n=1}^\infty n\left[\hat{F}^{(n)}(t) - \hat{F}^{(n+1)}(t)\right] = \sum_{n=1}^\infty \hat{F}^{(n+1)}(t) \geq \sum_{n=1}^\infty F^{(n+1)}(t) \text{ [from (4.5.2)]}$$

$$\sum_{n=1}^\infty n\left[F^{(n)}(t) - F^{(n+1)}(t)\right] = U(t) \Rightarrow \hat{U}(t) \geq U(t).$$

Theorem 8 $t/\mu - 1 \leq U(t) \leq t/\mu$.

Proof. We have $= t/\mu$ and $U(t) \geq t/\mu - 1$ by 4.5 and by theorem 4.5.1.
Also by theorem 7 we have $\hat{U}(t) \geq U(t)$.
Hence by combining the above results we get

$$\frac{t}{\mu} - 1 \leq U(t) \leq \frac{t}{\mu}. \tag{4.5.3}$$

Theorem 9 *If the c.d.f. corresponding to an exponential variate $G(t) = 1 - e^{-t/\mu}$, then*

$$\sum_{n=1}^\infty F^{(n)}(t) \leq \frac{t}{\mu} = \sum_{n=1}^\infty G^{(n)}(t),$$

where $F^{(n)}(t)$ represents the probability of at least n number of renewals up to t and $G^{(n)}(t)$ the same under a Poisson process with parameter μ under NBUE. *The inequality is reversed under* NWUE.

Proof. We have

$$U(t) = \sum_{n=1}^\infty nP_n(t) = \sum_{n=1}^\infty n\left[F^{(n)}(t) - F^{(n+1)}(t)\right] = \sum_{n=1}^\infty F^{(n)}(t) \leq \frac{t}{\mu} \text{ by (4.5.3)}$$

and $\sum_{n=1}^{\infty} G^{(n)}(t) =$ average number of renewals $(0, t]$ under a Poisson process with parameter μ is given by t/μ.

Hence, we have

$$\sum_{n=1}^{\infty} F^{(n)}(t) \leq \frac{t}{\mu} = \sum_{n=1}^{\infty} G^{(n)}(t). \tag{4.5.4}$$

Theorem 10 *Let the cumulative distribution function corresponding to an exponential variate*

$$\begin{aligned} G(t) &= 1 - e^{-t/\mu} \\ \textit{and } \bar{G}(t) &= 1 - G(t) = e^{-t/\mu}, \end{aligned}$$

then

$$\sum_{n=1}^{\infty} F^{(n)}(t) \leq \frac{t}{\mu} = \sum G^{(n)}(t)$$

holds if NBUE *holds for F and the sign of inequality is reversed if F is* NWUE.

Proof. We have

$$\begin{aligned} U(t) &= \sum_{n=1}^{\infty} nP_n(t) = \sum_{n=1}^{\infty} n\left[F^{(n)}(t) - F^{(n+1)}(t)\right] \\ &= \sum_{n=1}^{\infty} F^{(n)}(t), \end{aligned}$$

where $F^{(n)}(t)$ represents the probability of at least n renewals replacements up to time t and $P_n(t)$ represents the probability of exactly n number of replacements up to time t. Also the probability of n replacements in a Poisson process (which generates exponential distribution) is $(1/\mu)t$ and

$$\sum_{n=1}^{\infty} G^{(n)}(t) = \sum_{n=1}^{\infty} n\left[G^{(n)} - G^{(n+1)}(t)\right]$$

$=$ average number of replacements in a Poisson process with parameter $(1/\mu)$ is

$$\frac{1}{\mu}t = \hat{U}(t) \text{ (from 4.5.3)}$$

Again, since $\hat{U}(t) \geq U(t)$ for NBUE, therefore,

$$\sum_{n=1}^{\infty} F^{(n)}(t) \leq \frac{t}{\mu} = \sum_{n=1}^{\infty} G^{(n)}(t). \tag{4.5.5}$$

4.5.2 A result

If F_1s are NBUE, with mean $\mu (i = 1, 2, \ldots)$, then

$$\sum_{i=j}^{\infty} F_1 * F_2 * F_3 * \ldots F_i(t) \leq \sum_{n=j}^{\infty} G^{(n)}(t),$$

where $j = 1, 2, \ldots; t \geq 0$, $*$ indicates convolution and G is the cumulative distribution functions corresponding to an exponential variate with mean μ whose n-fold convolution is denoted by $G^{(n)}$.

Case I *Proof. First we prove that the result is true for* $j = 1$. *In particular if all* F_i*s are identical and equal to* F, *say, then the* L.H.S. *of the result (to be proved) becomes equal to* $\sum_{n=1}^{\infty} F^{(n)}(t)$ *which is* $\leq$ *the* R.H.S. $\sum_{n=1}^{\infty} G^{(n)}(t)$ *by theorem 9. Next suppose,* $F_k = F_{k+1} = \ldots$ *but the first* k *c.d.f.s, viz.,* $F_1, F_2, \ldots, F_k$ *differ.*

In this case, for $j = 1$, the L.H.S. is

$$\begin{aligned}
\sum_{I=1}^{\infty} F_1 * F_2 * \cdots * F_i(t) &= \int_0^t f_1(x)\,dx + \left[\int_0^t f_1(x) \sum_{j=1}^{\infty} F_2 * F_3 * \ldots F_{j+1}(t-x)\right] dx, \\
&\quad \text{where } \frac{d}{dx}F_1(x) = f_1(x) \\
&= \int_0^t \left[1 + \sum_{j=1}^{\infty} F_2 * F_3 * \cdots * F_{j+1}(t-x)\right] dF_1(x) \\
&= \int_0^t \left[1 + \sum_{j=1}^{\infty} F_2^{(j)}(t-x)\right] dF_1(x), \\
&\quad \text{where } F_2 = F_3 = \cdots = F_{j+1}, \text{ (say)} \\
&\leq \int_0^t \left[1 + \sum_{j=1}^{\infty} G^{(j)}(t-x)\right] dF_1(x) \text{ by theorem 9} \\
&= \int_0^t \left[1 + \frac{t-x}{\mu}\right] dF_1(x) \text{ by (4.5.4)} \\
&= F_1(t) + \frac{t}{\mu}F_1(t) - \frac{1}{\mu}\int_0^t x\,dF_1(x) \ (\because F_1(0) = 0) \\
&= F_1(t) + \frac{1}{\mu}\int_0^t F_1(x)\,dx = F_1(t) + \frac{1}{\mu}\int_0^t (1 - R_1(x))\,dx \\
&= F_1(t) + \frac{t}{\mu} - \frac{1}{\mu}\int_0^t R_1(x)\,dx, \text{ where } R_1(x) = 1 - F_1(x). \quad (4.5.6)
\end{aligned}$$

$$\left(\because -\frac{1}{\mu}\int_0^t x\,dF_1(x) = -\frac{1}{\mu}xF_1(x)\Big|_0^t + \frac{1}{\mu}\int_0^t F_1(x)\,dx = -\frac{1}{\mu}tF_1(t) + \frac{1}{\mu}\int_0^t F_1(x)\,dx\right).$$

Also, since F_1 is NBUE, therefore,

$$\frac{\int_t^\infty R_1(x)\,dx}{R_1(t)} \le \mu \;\Rightarrow\; \int_t^\infty R_1(x)\,dx \le \mu R_1(t)$$

$$\Rightarrow\; \mu - \int_t^\infty R_1(x)\,dx \ge \mu - \mu R_1(t) = \mu F_1(t)$$

$$\Rightarrow\; \int_0^t \frac{R_1(x)}{\mu}\,dx \ge F_1(t)$$

$$\Rightarrow\; F_1(t) \le \int_0^t \frac{R_1(x)}{\mu}\,dx. \tag{4.5.7}$$

Since, from (4.5.6)

$$\sum_{i=1}^{\infty} F_1 * F_2 * \cdots * F_i(t) \le F_1(t) + \frac{t}{\mu} - \frac{1}{\mu}\int_0^t R_1(x)\,dx$$

and from (4.5.7)

$$F_1(t) \le \int_0^t \frac{R_1(x)}{\mu}\,dx.$$

It follows that

$$\sum_{t=1}^{\infty} F_1 * F_2 * \cdots * F_i(t) \le \frac{t}{\mu}.$$

Also,

$$\sum_{n=1}^{\infty} G^{(n)}\,(t) = \frac{t}{\mu}.$$

Therefore,

$$\sum_{j=1}^{\infty} F_1 * F_2 * \cdots * F_i(t) \le \sum_{n=1}^{\infty} G^{(n)}\,(t) \text{ by taking } k \to \infty. \tag{4.5.8}$$

Case II *We give a proof of the result when $j > 1$, by induction. The result* (4.5.8) *holds for $j = 1$ by Case* I. *Assume that the result in general holds for $j = 1, 2, 3, \ldots, k$. Now to show that the result holds for $j = k + 1$.*

Proof. Consider

$$
\begin{aligned}
\sum_{i=k+1}^{\infty} F_1 * F_2 * \cdots * F_i(t) &= \sum_{i=k+1}^{\infty} \int_0^t [F_2 * \cdots * F_i(t-x)] f_1(x)\, dx \\
&= \sum_{i=k+1}^{\infty} \int_0^t [F_2 * \cdots * F_i(t-x)]\, dF_1(x) \\
&= \int_0^t \sum_{i=k+1}^{\infty} [F_2 * \cdots * F_i(t-x)]\, dF_1(x) \\
&= \int_0^t \sum_{i=k}^{\infty} [F_2 * \cdots * F_{i-1}\,(t-x)]\, dF_1(x) \\
&\leq \int_0^t \sum_{i=k}^{\infty} G^{(i)}\,(t-x)\, dF_1(x)
\end{aligned}
$$

From (4.5.7) on the hypothesis of induction, viz., the result holds for $i = 1, 2, \ldots, k$

$$
\begin{aligned}
&= \int_0^t \sum_{i=k}^{\infty} G^{(i)}\,(t-x) f_1(x)\, dx \\
&= \int_0^t \sum_{i=k}^{\infty} G^{(i)}\,(t-x)\, dF_1(x) \\
&= \int_0^t \sum_{i=k}^{\infty} (G^{(I-1)} * F_1)(t-x)\, dG(x) \\
&\leq \int_0^t \sum_{i=k}^{\infty} G^{(i)}\,(t-x)\, dG(x) \\
&= \sum_{i=k}^{\infty} G^{(i+1)}\,(t) = \sum_{i=k+1}^{\infty} G^{(i)}\,(t)
\end{aligned}
$$

$$
\Rightarrow \sum_{i=k+1}^{\infty} F_1 * F_2 * \cdots * F_i(t) \leq \sum_{i=k+1}^{\infty} G^{(i)}\,(t)
$$

which proves the result.

4.5.3 Corollary to theorem

If $0 \leq h(1) \leq h(3) \ldots$ and F is NBUE, then

$$
\sum_{n=1}^{\infty} h(n) F^{(n)}\,(t) \leq \sum_{n=1}^{\infty} h(n) G^{(n)}\,(t). \qquad (4.5.9)
$$

We have,

$$\begin{aligned}\sum_{n=1}^{\infty} h(n)F^{(n)}(t) &= h(1)F^{(1)}(t) + h(2)F^{(2)}(t) + \dots \\ &= h(1)\sum_{n=1}^{\infty} F^{(n)}(t) + (h(2) - h(1))\sum_{n=2}^{\infty} F^{(n)}(t) + (h(3) \\ &\quad -h(2))\sum_{n=3}^{\infty} F^{(n)}(t) + \dots \\ &\leq h(1)\sum_{n=1}^{\infty} G^{(n)}(t) + (h(2) - h(1))\sum_{n=2}^{\infty} G^{(n)}(t) + (h(3) \\ &\quad -h(2))\sum_{n=3}^{\infty} G^{(n)}(t) + \dots \\ &= \sum_{n=1}^{\infty} h(n)G^{(n)}(t).\end{aligned}$$

Theorem 11 *If F is* NBUE *and $c(k)$ is convex increasing them*

$$\sum_{k=1}^{\infty} c(k)P[N(t) = k] \leq \sum_{k=1}^{\infty} c(k)\frac{e^{-\lambda t}(\lambda t)^k}{k!} \dots \tag{4.5.10}$$

Proof.

$$\begin{aligned}\text{L.H.S.} &= \sum_{k=1}^{\infty} c(k)\left[F^{(k)}(t) - F^{(k+1)}(t)\right] \\ &= c(1)\left[F^{(1)}(t) - F^{(2)}(t)\right] + c(2)\left[F^{(2)}(t) - F^{(3)}(t)\right] \\ &\quad + c(3)\left[F^{(3)}(t) - F^{(4)}(t)\right] + \dots \\ &= F^{(1)}(t)(c(1) - c(0)) + F^{(2)}(t)(c(2) - c(1)) \\ &\quad + F^{(3)}(t)(c(3) - c(2)) + \dots\end{aligned}$$

Since, $c(k)$ is convex increasing we have,

$$c(k) - c(k-1) = h(k)$$

is an increasing function of k. Therefore,

$$\begin{aligned}\text{L.H.S.} &= \sum_{k=1}^{\infty} (c(k) - c(k-1))F^{(k)}(t) \\ &= \sum_{k=1}^{\infty} h(k)F^{(k)}(t) \leq \sum_{k=1}^{\infty} h(k)G^{(k)}(t), \quad \text{by (4.5.9)}\end{aligned}$$

Also

$$\begin{aligned}
\sum_{k=1}^{\infty} h(k)G^{(k)}(t) &= \sum_{k=1}^{\infty} (c(k)-c(k-1))G^{(k)}(t) \\
&= \sum_{k=1}^{\infty} c(k)(G^{(k)}(t)-G^{(k+1)}(t)) \\
&= \sum_{k=1}^{\infty} c(k)P[N(t)=k \mid N(t) \quad \text{[is a Poisson counting process with parameter } \lambda t.] \\
&= \sum_{k=1}^{\infty} c(k)\frac{e^{-\lambda t}}{k!}(\lambda t)^k.
\end{aligned}$$

Hence the theorem is proved.

An application of the result of Theorem 11: Let k be the # of spare parts demanded and N be the same in stock. Then for $k \geq N$, $(k-N)$ = shortage. If $c(k)$ represents the shortage, then

$$c(k) = \begin{cases} k-N, & \text{for } k \geq N \\ 0, & \text{for } k < N. \end{cases}$$

$c(k)$ is obviously convex on its domain of definition.

Expected shortage is given by

$$\begin{aligned}
\Sigma c(k)P(N_t = k) &= \Sigma c(k)\left[F^{(k)}(t) - F^{(k+1)}(t)\right] \\
&= \Sigma\left[c(k) - c(k-1)\right]F^{(k)}(t) \\
&\leq \Sigma\left[c(k) - c(k-1)\right]G^{(k)}(t) \\
&= \Sigma c(k)\left[G^{(k)}(t) - G^{(k+1)}(t)\right]. \qquad (4.5.11)
\end{aligned}$$

Therefore, the upper bound of the expected shortage

$$\begin{aligned}
&= \sum_{k-N}^{\infty} c(k)\frac{e^{-\lambda t}(\lambda t)^k}{k!} = \sum_{k=N}^{\infty}(k-N)\frac{e^{-\lambda t}(\lambda k)^k}{k!} \\
&= \sum_{k=N}^{\infty} k\frac{e^{-\lambda t}(\lambda t)^k}{k!} - N\sum_{k=N}^{\infty}\frac{e^{-\lambda t}(\lambda t)^k}{k!} \\
&= \lambda t \sum_{k=N-1}^{\infty}\frac{e^{-\lambda t}(\lambda t)^k}{k!} - N\sum_{k=N}^{\infty}\frac{e^{-\lambda t}(\lambda t)^k}{k!}.
\end{aligned}$$

Again,

$$\sum_{k=N-1}^{\infty}\frac{e^{-\lambda t}(\lambda t)^k}{k!} = \frac{1}{\Gamma(N-1)}\int_0^{\lambda t} e^{-\xi}\xi^{N-1}\,d\xi$$

and similarly

$$\sum_{k=N}^{\infty}\frac{e^{-\lambda t}(\lambda t)^k}{k!} = \frac{1}{\Gamma(N)}\int_0^{\lambda t} e^{-\xi}\xi^{N}\,d\xi.$$

$\therefore$ u.b. expected shortage

$$= \frac{N\lambda t}{\Gamma(n)} \int_0^{\lambda t} e^{-\xi} \xi^{N-1} \, d\xi - \frac{N}{\Gamma(N)} \int_0^{\lambda t} e^{-\xi} \xi^{N} \, d\xi. \tag{4.5.12}$$

4.5.4 Availability

Let

$$X(t) = \begin{cases} 1, & \text{the system is operating at time } t \\ 0, & \text{otherwise.} \end{cases}$$

Then

$$P[X(t) = 1] = A(t),$$

where $A(t)$ is called *availability* at any time t.

$$\begin{aligned} E[X(t)] &= P[X(t) = 1] \cdot 1 + P[X(t) = 0] \cdot 0 \\ &= P[X(t) = 1] = A(t). \end{aligned}$$

Limiting Availability: If the $\lim A(t)$ exists as $t \to \infty$ and is equal to A (*say*), then A is called the *Limiting Availability* of the system.

Average Availability: Average availability in $(0, T]$ is defined as

$$\int_0^T A(t) \frac{dt}{T}. \tag{4.5.13}$$

Limiting Average Availability: It is defined as

$$= \lim_{T \to \infty} \left[\frac{1}{T} \int_0^T A(t) \, dt \right]. \tag{4.5.14}$$

NOTE

If repairs are not permitted then A(t) reduces to system reliability, i.e., the probability that the system operates without failure during (0, t].

Let $O(t)$ be the total amount of operating time in $(0, t]$. Then,

$$O(T) = \int_0^T X(t) \, dt$$

$$\Rightarrow \frac{O(T)}{T} = \frac{1}{T} E \left[\int_0^T X(t) \, dt \right] = \frac{1}{T} \int_0^T E(X(t)) \, dt = \frac{1}{T} \int_0^T A(t) \, dt.$$

By Fubini's theorem giving that each of the above integrals exists.

A Result: Limiting average availability is the limiting availability.

Proof. We require to show that

$$\lim_{T \to \infty} \frac{1}{T} \int_0^T A(t) \, dt = \lim_{t \to \infty} \lim A(t) = A$$

$$\lim_{t\to\infty} A(t) = A \Rightarrow |A(t) - A| < \epsilon \forall t > t_0 \Rightarrow -\epsilon < A(t) < A + \epsilon \forall t > t_0$$

$$\therefore \quad \frac{1}{T}\int_0^T A(t)\,dt = \frac{1}{T}\int_0^{t_0} A(t)\,dt + \frac{1}{T}\int_{t_0}^T A(t)\,dt < \frac{1}{T}\int_0^{t_0} A(t)\,dt + \frac{1}{T}\int_{t_0}^T (A+\epsilon)\,dt.$$

Since $t_0 < T$ and t_0 is fixed, therefore,

$$\lim_{T\to\infty}\int_0^T A(t)\,dt \le 0 \lim_{T\to\infty}\frac{(A+\epsilon)(T-t_0)}{T} = (A+\epsilon).$$

Similarly,

$$\lim_{T\to\infty}\int_0^T A(t)\,dt \ge (A+\epsilon).$$

Hence,

$$\lim_{T\to\infty}\int_0^T A(t)\,dt = A. \tag{4.5.15}$$

Next, we express limiting availability in terms of mean operation period and the mean repairing period. Let T_i be the duration of the ith operating period and D_i be the down time for the ith repair or replacement. While not assuming independence of T_i and D_i, we assume the independence of $(T_i + D_i)$ for $i = 1, 2, \ldots$ Let F be the common distribution of T_i and H be the common distribution of $T_i + D_i$ for $i = 1, 2, \ldots$ Then, the renewal function corresponding to H is given by

$$U_H(t) = \sum_{n=1}^{\infty} H^{(n)}(t).$$

Then the availability at any point of time t is either the reliability up to time t given by $R(t)$ (in case of no replacement or renewal up to time t) or the probability of survival up to time $t - \tau$, i.e., $R(t-\tau)$ and then a replacement at time τ given by $u(\tau)$ where τ ranges from 0 to t.

In other words,

$$\begin{aligned} A(t) &= R(t) + \int_0^t R(t-\tau)u(\tau)\,d\tau, \\ \text{or, } A(t) &= R(t) + \int_0^t R(t-\tau)\,dU(\tau) \end{aligned}$$

in terms of renewal density or renewal function.

By the Key Renewal Theorem of Smith (*vide* Theorem 4), we have

$$\lim_{t\to\infty} A(t) = \frac{E(t)}{E(T)+E(D)}, \tag{4.5.16}$$

where T and D are uptime and downtime respectively when H is a non-lattice distribution.

4.5.5 Maintenance through spares and repairs

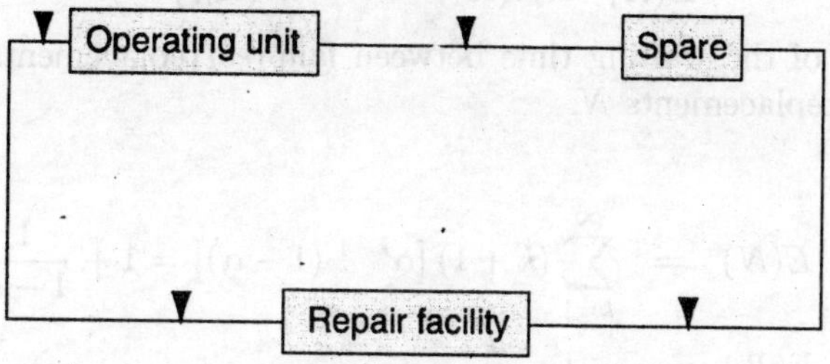

Fig 22

We present three models.

Model I deals with a single unit system with a single standby with repair facility. *Model* II (dealt in Section 4.6) is a generalisation of Model I based on exponential failure and repair distribution. Finally, *Model* III deals with the reliability of a system supplied by the spares (*vide* section 4.6).

Model I *As stated earlier we consider here a single unit system with a single standby and repair facility. When the operating unit fails it is replaced by the spare in negligible time. At the same time the failed unit is sent for repair. This instant is considered as a point of regeneration. There is system failure when the operating component fails while the spare is still under repair (i.e., the repair of the previously failed unit is yet to be completed).*

Let the operating unit has the c.d.f. of life as F with mean μ and that of repair time has the c.d.f. as G with mean ν. Assume that the two distributions are independent. Finally, assume that at $t = 0$, the operating system and standby are new. The maintenance cycle is shown in Fig 22.

We denote the time for the first system failure as

$$T_1 = \tilde{X}_1 + X_2 + \cdots + X_n,$$

where $\tilde{X}_1$ represents the failure time of the first failure which is replaced by a spare with probability unity. X_is $(i = 2, 3, \ldots, N)$ represent the failure time of the ith component which may be replaced with sparc with probabillty α. $\tilde{X}_i$ and X_i $(i = 2, 3, \ldots, N)$ are identically distributed r.v.s. following a c.d.f. FN represent the number of operating unit failures culminating to a system failure.

Clearly, N has a modified Geometric distribution given by

$$P[N = k + 1] = 1 \cdot \alpha^{k-1}\,(1 - \alpha),$$

where

$$\alpha = \int_0^t G(t) f(t)\,dt = \int_0^t G(t)\,dF(t) \text{ and } f(t) = \frac{d}{dt}F(t),$$

F and G being the c.d.f. of failure distribution and the repairment respectively.

Now

$$E(T_1) = E(X_i)E(N) = \mu E(N).$$

Since

$$E(\tilde{X}_1) = E(X_2) = \cdots = E(X_N) = \mu$$

and the distribution of the waiting time between failures/replacements being independent of the total number of replacements N.

Now,

$$\begin{aligned} E(N) &= \sum_{k=1}^{\infty} (k+1)\left[\alpha^{k-1}(1-\alpha)\right] = 1 + \frac{1}{1-\alpha} \\ E(T_1) &= \mu + \frac{\mu}{1-\alpha}. \end{aligned}$$

If T is the time of system failure measured from the first regeneration point, i.e., $T = X_2 + \cdots + X_N$, then

$$E(T) = \mu/(1-\alpha).$$

Limiting availability of the system is given by

$$\lim_{t\to\infty} A(t) = \frac{E(T)}{[E(T)+E(D)]},$$

where T and D are uptime and downtime respectively.

Therefore,

$$E(T) = \mu/1-\alpha.$$

However, to get $E(D)$, we require to note the system. System failure occurs when the operating system fails before the system undergoing the repair is returned to service as a spare (i.e., following completion of a repair). Let the failure of the operating system after a time x occurs with probability $f(x)\,dx = dF(x)$. Again the given system under failure condition at time x and not being replaced then occurs with probability $\bar{G}(x)$, the conditional probability that it will not be repaired till a further period t, i.e., up to $(t+x)$ is given by

$$\frac{\bar{G}(t+x)}{\bar{G}(x)}, \quad \text{where } \bar{G}(x) = 1 - G(x), \quad 0 \le x < \infty, \quad 0 \le t < \infty.$$

Therefore, the expected downtime is given by

$$E(D) = \int_0^{\infty}\int_0^{\infty} \frac{\bar{G}(t+x)}{\bar{G}(x)}\, dt dF(x).$$

Hence the limiting availability is given by

$$\frac{\mu(1-\alpha)^{-1}}{\mu(1-\alpha)^{-1} + \displaystyle\int_0^{\infty}\int_0^{\infty} \frac{\bar{G}(t+x)}{\bar{G}(x)}\, dt\, dF(x)}. \tag{4.5.17}$$

▶▶ NOTE

Limiting system availability depends on the failure and repair distributions, viz., F and G respectively; and not on their means.

4.6 A Formula for the Distribution of the Time of First System Failure

Let H be the distribution of time of first system failure from start and H_1 be the distribution of time from first regeneration point to system failure.

Then

$$H(t) = \int_0^t H_1(t-x)f(x)\,dx = \int_0^t H_1(t-x)\,dF(x). \tag{4.6.1}$$

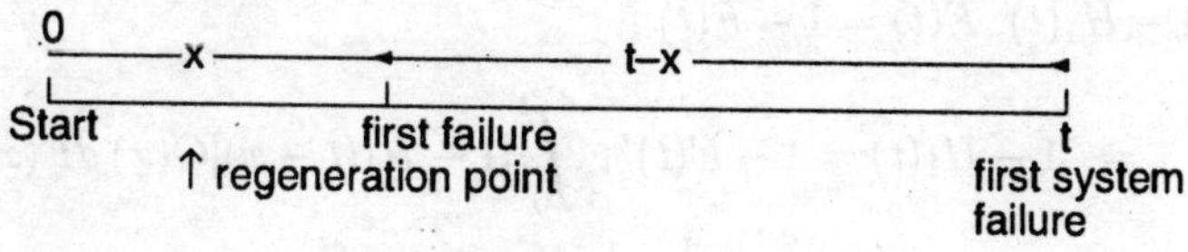

Fig 23

We express H_1 in terms of H and G using the following integral equation, viz.

$$\begin{aligned}
\bar{H}_1(t) &= \bar{F}(t) + \int_0^t \bar{H}_1(t-x)G(x)\,dF(x)\\
\Rightarrow 1 - H_1(t) &= 1 - F(t) + \int_0^t (1 - H_1(t-x))G(x)\,dF(x)\\
\Rightarrow H_1(t) &= F(t) - \int_0^t G(x)\,dF(x) + \int_0^t H_1(t-x)G(x)\,dF(x).
\end{aligned}$$

Taking total differential on both sides of (4.6.1) we have,

$$dH_1(t) = dF(t) - G(t)\,dF(t) + (dH_1) * (GdF(t)).$$

Taking L-S transformation on both sides, we have

$$\begin{aligned}
\int_0^\infty e^{-st}\,dH_1(t) &= \int_0^\infty e^{-st}\,dF(t) - \int_0^\infty e^{-st}G(t)\,dF(t)\\
&\quad + \int_0^\infty e^{-st}G(t)\,dF(t)\int_0^\infty e^{-st}\,dH_1(t)\\
\Rightarrow H_1^*(s) &= F^*(s) - k^*(s) + k^*(s)H_1^*(s)\\
\Rightarrow H_1^*(s)(1-k^*(s)) &= F^*(s) - k^*(s)\\
\Rightarrow H_1^*(s) &= \frac{F^*(s) - k^*(s)}{1 - k^*(s)}.
\end{aligned} \tag{4.6.2}$$

Again from (4.6.1),

$$\begin{aligned}
H(t) = \int_0^t H_1(t-x)\,dF(x) \;&\Rightarrow\; H_1^*F\\
&\Rightarrow\; dH(t) = d(H_1(t)^*F(t))\\
&\Rightarrow\; \int_0^\infty e^{-st}\,dH(t) = \int_0^\infty e^{-st}\,dH_1(t)\int_0^\infty e^{-st}\,dF(t)\\
&\Rightarrow\; H^*(s) = H_1^*(s)F^*(s).
\end{aligned}$$

Therefore,

$$H^*(s) = F^*(s)\frac{F^*(s) - k^*(s)}{1 - k^*(s)}. \tag{4.6.3}$$

By taking inverse Laplace transformation on both sides we can get $H(t)$ in terms of $F(t)$ and $k(t)$.

$$\bar{H}_1(t) = \bar{F}(t) + \int_0^t \bar{H}_1(t-x)G(x)\,dF(x),$$

where $\bar{H}_1(t) = 1 - H_1(t)$, $\bar{F}(t) = 1 - F(t)$

$$\Rightarrow 1 - H_1(t) = 1 - F(t) + \int_0^t [1 - H_1(t-x)]G(x)\,dF(x)$$

$$\Rightarrow H_1(t) = F(t) - \int_0^t G(x)\,dF(x) + \int_0^t H_1(t-x)G(x)\,dF(x). \tag{4.6.4}$$

Model II *This is a generalisation of the earlier model, based on exponential failure and exponential repairment distribution. It may be noted that in the general case, i.e., in the case of arbitrary failure and repairment distribution, a generalisation of the earlier model (i.e., with, say, m operating units, n spares and s service facilities) is confronted with several difficulties and the mathematical analysis becomes very much complicated because no regeneration point exists. That is why the reason of generalising the earlier model with exponential distribution.*

We assume m operating units, n spares and s service facilities. Each operating unit has a life distribution given by

$$F(t) = 1 - e^{-\lambda t}$$

and the repair time distribution

$$G(t) = 1 - e^{-\gamma t}.$$

A unit in spare status cannot fail. All life lengths and repair times are mutually independent random variables. As in model I, when an operating unit fails, it is immediately replaced by a spare, if available. The failed unit is immediately sent to the repair facility. Repair begins immediately unless all the s service channels are engaged.

Let $X(t)$ be the number of units down at time t (i.e., the said units are either under repair or awaiting repair. Then,

$$\{X(t), 0 \le t < \infty\}$$

is a Markov chain on continuous time. That is, for continuous time points $0 \le t_1 < t_2 \cdots < t_k$ and discrete (non-negative integers) quantities $x_1, x_2, \ldots, x_k$, we have

$$\begin{aligned} P\{X(t_k) &= x_k \mid X(t_1) = x_1,\ X(t_2) = x_2, \ldots, X(t_{k-1}) = x_{k-1}\} \\ = P\{X(t_k) &\doteq x_k \mid X(t_{k-1}) = x_{k-1}\}. \end{aligned} \tag{4.6.5}$$

Also, the steady state probability is given by

$$\lim_{t_i \to \infty} P(X(t_i)) = \pi_i,\ i = 1, 2, \ldots, k.$$

The Kolmogorov equations under the set-up (i.e., with one operating unit, one spare and one repair facility) is given by

$$p_0(t + \delta t) = p_0(t)[1 - \lambda\delta t + o(\delta t))] + p_1(t)[\mu\delta t + o(\delta t)]. \tag{4.6.6}$$

$$(4.6.6) \Rightarrow p_0(t) = -\lambda p_0(t) + \mu p_1(t) \tag{4.6.7}$$

$$p_1(t+\delta t) = p_0(t)[\lambda \delta t + o(\delta t)] + p_1(t)[1-(\lambda+\mu)\delta t + o(\delta t)] + p_2(t)[\mu \delta t + o(\delta t)] \tag{4.6.8}$$

$$(4.6.3) \Rightarrow p_1(t) = \lambda p_0(t) - (\lambda+\mu)p_1(t) + \mu p_2(t). \tag{4.6.9}$$

Now it can be shown that

$$\lim_{t\to\infty} p_i(t) = 0$$

$$\text{and} \quad \lim_{t\to\infty} p_i(t) = \pi_i, \ \forall\, i = 1, 2, \ldots$$

Therefore, (4.6.8) and (4.6.9) → to the following as $t \to \infty$, viz.

$$-\lambda \pi_0 + \mu \pi_1 = 0 \tag{4.6.10}$$

$$\text{and} \quad \lambda \pi_0 - (\lambda+\mu)\pi_1 + \mu \pi_2 = 0. \tag{4.6.11}$$

Also,

$$\pi_0 + \pi_1 + \pi_2 = 1. \tag{4.6.12}$$

Solving (4.6.10), (4.6.11) and (4.6.12), by Crammer's method, we have

$$\pi_0 = \frac{\begin{vmatrix} 0 & \mu & 0 \\ 0 & -(\lambda+\mu) & \mu \\ 1 & 1 & 1 \end{vmatrix}}{\begin{vmatrix} -\lambda & \mu & 0 \\ \lambda & -(\lambda+\mu) & \mu \\ 1 & 1 & 1 \end{vmatrix}}. \tag{4.6.13}$$

$$\pi_1 = \frac{\begin{vmatrix} -\lambda & 0 & \mu \\ \lambda & 0 & \mu \\ 1 & 1 & 1 \end{vmatrix}}{\begin{vmatrix} -\lambda & \mu & 0 \\ \lambda & -(\lambda+\mu) & \mu \\ 1 & 1 & 1 \end{vmatrix}} \tag{4.6.14}$$

and

$$\pi_2 = \frac{\begin{vmatrix} -\lambda & \mu & 0 \\ \lambda & -(\lambda+\mu) & 0 \\ 1 & 1 & 1 \end{vmatrix}}{\begin{vmatrix} -\lambda & \mu & 0 \\ \lambda & -(\lambda+\mu) & \mu \\ 1 & 1 & 1 \end{vmatrix}} \tag{4.6.15}$$

On expansion we get,

$$\pi_0 = \left(1 + \frac{\lambda}{\mu} + \frac{\lambda^2}{\mu^2}\right)^{-1}. \tag{4.6.16}$$

$$\pi_1 = \frac{\lambda}{\mu}\left(1 + \frac{\lambda}{\mu} + \frac{\lambda^2}{\mu^2}\right)^{-1}. \tag{4.6.17}$$

and

$$\pi_2 = \frac{\lambda^2}{\mu^2}\left(1 + \frac{\lambda}{\mu} + \frac{\lambda^2}{\mu^2}\right)^{-1}. \tag{4.6.18}$$

Limiting availability

$$\pi_0 + \pi_1 = \left(1 + \frac{\lambda}{\mu}\right)\left(1 + \frac{\lambda}{\mu} + \frac{\lambda^2}{\mu^2}\right)^{-1}. \tag{4.6.19}$$

Model III (A spare inventory model) *Hitherto we have been considering reliability of a system supplied by spares. In the present model, we consider the operation of supplying the spares.*

Let at time $t = 0$, maintenance depot has n spares available of a given component. Customers bring failed components to the depot at times governed by a Poisson process with rate λ. A failed unit is replaced by a spare, if available. Repair of a failed unit starts immediately as the number of repairment channel is unlimited. The time for repair follows a distribution G (cumulative distribution function). Also repairment of components are made independently and are independent of the arrival time of customer. Finally, the repaired components become part of the spares inventory.

Theorem 12 *The probability of the number of components undergoing repair at any time, say, $N(t)$ is given by*

$$P[N(t) = j] = \frac{\left(\lambda \int_0^t \bar{G}(\tau)\, d\tau\right)^j}{j!} \cdot e^{-\lambda\left[\int_0^t \bar{G}(\tau)\, d\tau\right]^j}$$

$$\text{and } \lim_{t\to\infty} P(N(t) = j) = e^{-\lambda\nu}\frac{(\lambda\nu)^j}{j!},$$

where

$$\nu = \int_0^\infty \bar{G}(\tau)\, d\tau \text{ and } \bar{G}(t) = 1 - G(t). \tag{4.6.20}$$

At the Poisson rate λ, the probability of m # of demands for spares in time t is $e^{-\lambda t}\frac{(\lambda t)^m}{m!}$. Given that there are j# of demands then, the range of m is $[j, \infty)$. Again if $P(t)$ is the probability of a component not getting repaired at any time t, then the probability of j components not getting repaired at time t out of m is

$$\binom{m}{j}\ (P(t))^j(1 - P(t))^{m-j}.$$

The simultaneous probability of demand of spares being m and j components not being repaired is

$$e^{-\lambda t}\frac{(\lambda t)^m}{m!}\cdot \binom{m}{j}\ (P(t))^j(1 - P(t))^{m-j}.$$

For any value of m lying between j to infinity is

$$\sum_{m=j}^{\infty} e^{-\lambda t}\frac{(\lambda t)^m}{m!}\binom{m}{j}(P(t))^j(1-P(t))^{m-j}$$

$$= \sum_{m=j}^{\infty} e^{-\lambda t}\frac{(\lambda t)^j}{j!}\frac{(\lambda t)^{m-j}}{(m-j)!}\left(\frac{j!(m-j)!}{m!}\frac{m!}{j!(m-j)!}\right)(P(t))^j(1-P(t))^{m-j}$$

$$= \sum_{m=j}^{\infty} e^{-\lambda t}\frac{(\lambda t)^{m-j}}{(m-j)!}(1-P(t))^{m-j}\frac{(\lambda t)^j}{j!}(P(t))^j$$

$$= e^{-\lambda t}\left\{\sum_{m-j=0}^{\infty}\frac{[\lambda t(1-P(t)]^{m-j}}{(m-j)!}\right\}\frac{(\lambda t)^j}{j!}[P(t)]^j$$

$$= e^{-\lambda t}e^{\lambda t(1-P(t))}\frac{[\lambda t(P(t)]^j}{j!} = e^{-\lambda t P(t)}\frac{[\lambda t(P(t)]^j}{j!}. \tag{4.6.21}$$

Finally, we require to obtain $P(t)$ and that has not been repaired up to time t.

Now $P(t)$ is the probability that a typically failed unit brought at a time between x and $(x+dx)$ and $0 \le x \le t$.

Now the probability of bringing a unit between $(x, x+dx)$ is $\frac{dx}{t}$ and the probability that it has not been repaired up to t, i.e., during the period $(t-x)$ is $\bar{G}(t-x)$. Therefore, the simultaneous probability of materialisation of both when x lies between $[0,t]$ is

$$P(t) = \int_0^t \bar{G}(t-x)\frac{dx}{t} = \frac{1}{t}\bar{G}(u)\,du.$$

On substitution of the value $P(t)$ the result follows.

Therefore, on substitution of the value of $P(t)$ in (4.6.21), we have $P[N(t)=j]$ is

$$= e^{-\lambda t\cdot\frac{1}{t}\int_0^t \bar{G}(u)\,du}\frac{\left[\lambda t\frac{1}{t}\int_0^t \bar{G}(u)\,du\right]^j}{j!} = \frac{e^{-\lambda\int_0^t \bar{G}(u)\,du}}{j!}\left[\lambda\int_0^t \bar{G}(u)\,du\right]^j. \tag{4.6.22}$$

4.6.1 A formula for the distribution of the time of first system failure

Let H be the distribution of the time of first system failure from start and H_1 be the distribution of time from first regeneration time to system failure. Then,

$$H(t) = \int_0^t H_1(t-x)f(x)\,dx = \int_0^t H_1(t-x)\,dF(x). \tag{4.6.23}$$

We express H_1 in terms of H and G using the following integral equation, viz.,

$$\bar{H}_1(t) = \bar{F}(t) + \int_0^t \bar{H}_1(t-x)G(x)\,dF(x) - \int_0^t G(x)\,dF(x) \tag{4.6.24}$$

$$\Rightarrow 1 - H_1(t) = 1 - F(t) + \int_0^t (1 - H_1(t-x))G(x)\,dF(x)$$

$$\Rightarrow H_1(t) = F(t) - \int_0^t G(x)\,dF(x) + \int_0^t H_1(t-x)G(x)\,dF(x). \tag{4.6.25}$$

Taking total differential on both sides of (4.6.1), we have

$$dH_1(t) = dF(t) - G(t)\,dF(t) + (dH_1) * (GdF(t)).$$

Taking LS transformation on both sides, we have

$$\begin{aligned}\int_0^\infty e^{-st}\,dH_1(t) &= \int_0^\infty e^{-st}\,dF(t) - \int_0^\infty e^{-st}G(t)\,dF(t)\\ &\quad + \int_0^\infty e^{-st}G(t)\,dF(t)\int_0^\infty e^{-st}\,dH_1(t)\\ \Rightarrow H_1*(s) &= F*(s) - k*(s) + k*(s)H_1*(s),\end{aligned}$$

where

$$k*(s) = \int_0^\infty e^{-st}G(t)\,dF(t)$$

$$\text{and } H_1*(s) = \int_0^\infty e^{-st}\,dH_1(t).$$

$$\Rightarrow \quad H_1*(s)(1-k*(s)) = F*(s) - k*(s)$$

$$\Rightarrow \quad H_1*(s) = \frac{F*(s) - k*(s)}{(1-k*(s))} \tag{4.6.26}$$

Again since from (4.6.23)

$$\begin{aligned}& H(t) = \int_0^t H_1(t-x)f(x)\,dx = \int_0^t H_1(t-x)\,dF(x)) = H_1 * F\\ \Rightarrow\ & dH(t) = d(H_1(t) * F(t))\\ \Rightarrow\ & \int_0^\infty e^{-st}\,dH(t) = \int_0^t e^{-st}\,dH_1(t)\int_0^t e^{-st}\,dF(t)\\ \Rightarrow\ & H*(s) = H_1*(s)F*(s).\end{aligned}$$

Therefore,

$$H*(s) = F*(s)\frac{F*(s) - k*(s)}{(1-k*(s))} \text{ using (4.6.26)} \tag{4.6.27}$$

By taking inverse transformation on both sides we can get $H(t)$ in terms of $F(t)$ and $k(t)$.

Chapter 5

Coherent System

The basic motivation of this chapter is to investigate the structural relationship between a system and its components. We have seen earlier That, in a parallel system, the system works if any of the components works. In the same way, in a series system, the system works only if all the components work simultaneously. In general, most of the systems work Under mixed set up (i.e., k out of $n; k < n$) which implies that at least k components out of n must survive in order that the system works. A system becomes series if $k = n$ and parallel if $k = 1$.

Notations:

Consider the following notations in respect of an n-component system denoted by $X = (x_1, x_2, \ldots, x_n)$.

We define

$$x_i = \begin{cases} 1, & \text{if the } i\text{th component is surviving} \\ 0, & \text{otherwise.} \end{cases}$$

A structure function

$$\phi(X) = \phi(x_1, x_2, \ldots, x_n) = 1 \text{ or } 0$$

according as the system is working or not.

The set theoretic notations, viz.,

- $x_1 \cap x_2 \Rightarrow$ both x_1 and x_2 and
- $x_1 \cup x_2 \Rightarrow$ either of x_1 and x_2

are used to denote the series and the parallel bi-component system.

For an n-component parallel system the structure function given by

$$\phi(X) = \max(x_1, x_2, \ldots, x_n) = 1 - \prod_{i=1}^{n}(1 - x_i), \text{ since } x_i = 0, 1.$$

Also, the $\max(x_1, x_2, \ldots, x_n)$ is denoted as $\coprod_{i=1}^{n} x_i$ which is $1 - \prod_{i=1}^{n}(1 - x_i)$.

Similarly, for an n-component series system

$$\phi(X) = \min(x_1, x_2, \ldots, x_n) = \prod_{i=1}^{n} x_i, \text{ since } x_i = 0, 1.$$

Finally, for a k out of n structure, the structure function is represented as

$$\max[(x_1, x_2, \ldots, x_k)(x_1, x_2, \ldots, x_{k-i}, x_{k+1}), \ldots, (x_{n-k+1}, \ldots, x_n)] \text{ for } 1 \le k \le n.$$

5.1 Structure Function of a Hi-Fi System

Notation:

$x_1 \forall x_2 \Rightarrow$ either of x_1 or x_2

$x_1, x_2 \Rightarrow$ both x_1 and x_2.

(*i*) ***Structure function of a Hi-Fi system:*** A Hi-Fi stereo system has got the following components—(i) FM tuner, (*ii*) Record-player, (*iii*) Audio, (*iv*) Speaker I, (*v*) Speaker II.

The system is considered to be functioning if we can get the music either by FM tuner or by records. Also to get the music either of the two speakers must necessarily work. The system is diagrammatically illustrated below.

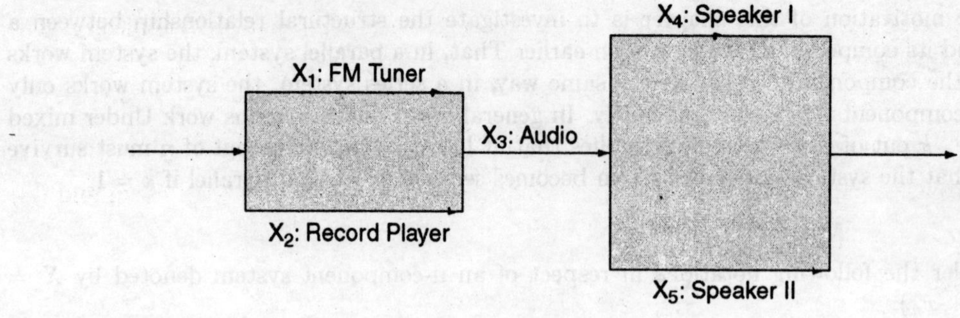

Fig 24

The structure function is given by

$$
\begin{aligned}
& (x_1 \cup x_2) \cap x_3 \cap (x_4 \cup x_5) \\
= \; & \coprod_{i=1}^{2} x_i \cdot (x_3) \cdot \coprod_{i=4}^{5} x_i = \left[1 - \prod_{i=1}^{2} (1 - x_i)\right] \cdot x_3 \left[1 - \prod_{i=4}^{5} 1 - (1 - x_i)\right] \\
= \; & [1 - (1 - x_1)(1 - x_2)] \cdot x_3[1 - (1 - x_4)(1 - x_5)] \\
= \; & x_1x_3x_4 + x_1x_3x_5 - x_1x_4x_5 + x_2x_3x_4 + x_2x_3x_5 - x_2x_4x_5 \\
& -x_1x_2x_3x_4 - x_1x_2x_3x_5 + x_1x_2x_3x_4x_5.
\end{aligned}
$$

(*ii*) ***Structure function of a four-engine aircraft where at least one engine on either side must necessarily work:*** E_1 and E_2 are two engines on one side of the aircraft, and E_3 and E_4 are the same on the other side of the aircraft. Their survival or otherwise are given by $x_i = 1$ or $x_i = 0$ $(i = 1, 2, 3, 4)$.

The structure function is given by

$$
\begin{aligned}
& (x_1 \cap x_3) \cup (x_1 \cap x_4) \cup (x_2 \cap x_3) \cup (x_2 \cap x_4) \\
= \; & (x_1 \cdot x_3) \cup (x_1 \cdot x_4) \cup (x_3 \cdot x_3) \cup (x_2 \cdot x_4) \\
= \; & [1 - (1 - x_1x_3)(1 - x_1x_4)(1 - x_2x_3)(1 - x_2x_4)] \\
= \; & x_1x_2x_3x_4 + x_1x_2x_3(1 - x_4) + x_1x_2x_4(1 - x_3) + x_2x_3x_4(1 - x_1) \\
& +x_1x_3(1 - x_2)(1 - x_4) + x_2x_3(1 - x_1)(1 - x_4) + x_1x_4(1 - x_2)(1 - x_3) \\
& +x_2x_4(1 - x_1)(1 - x_3).
\end{aligned}
$$

5.2 A Component Relevant to the Structure

If

$$\begin{aligned}\phi(x_1, x_2, x_3, \ldots, x_i = 1, \ldots, x_n) &= \phi(1_i, \bar{x}) = \phi(x_1, x_2, \ldots, x_i = 0, \ldots, x_n) \\ &= \phi(0_i, \bar{x}) = \phi(x_1, x_2, \ldots, x_{i-1}, x_{i+1}, \ldots, x_n) \\ &= \phi(x_1, x_2, \ldots, x_{i-1}, \ldots, x_{i+1}, \ldots, x_n),\end{aligned}$$

then the ith component is said to be irrelevant to the structure.

Symbolically,

$$\phi(1_i, \bar{x}) = \phi(0_i, \bar{x}) = \phi(1_i, \bar{x}). \tag{5.2.1}$$

If $\phi(\bar{x})$ does not decrease when x_is are increasing, then $\phi(\bar{x})$ is a monotonically non-decreasing function of $\bar{x}$.

5.3 Coherent Structure

A structure is said to be coherent if (i) each component x_i is relevant to the system and (ii) $\phi(\bar{x})$ is a non-decreasing function of $\bar{x}$.

5.4 A Lemma

$$\phi(\bar{x}) = x_i \phi(1_i, \bar{x}) + (1 - x_i)\phi(0_i, \bar{x}). \tag{5.4.1}$$

Proof. Put $x_i = 1$ in the R.H.S. of (5.4.1) $\Rightarrow x_i\phi(1_i, \bar{x}) = 1 \cdot \phi(x_1, x_2, \ldots, x_i = 1, \ldots, x_n) = \phi(\bar{x}) =$ L.H.S.

Put $x_i = 0$ in the R.H.S. $\Rightarrow (1 - x_i)\phi(0_i, \bar{x}) = \phi(x_1, x_2, \ldots, x_i = 0, \ldots, x_n) = \phi(\bar{x}) =$ L.H.S.

Thus the Lemma is proved.

5.5 A Result of Coherent Structure

$$\prod_{i=1}^{n} x_i \le \phi(\bar{x}) \le \coprod_{i=1}^{n} x_i, \tag{5.5.1}$$

where

$$\prod x_i = \min(x_1, x_2, \ldots, x_n)$$

and

$$\coprod x_i = \max(x_1, x_2, \ldots, x_n)$$

holds for a coherent structure.

Proof. Let

$$\prod_{i=1}^{n} x_i = \min(x_1, x_2, \ldots, x_n) = 1$$

$$\Rightarrow \quad x_1 = x_2 = \cdots = x_n = 1 \Rightarrow \phi(\bar{x}) = 1, \text{ where } \bar{x} \equiv (x_1, x_2, \ldots, x_n).$$

Hence the L.H.S. of the inequality holds.
Again, suppose

$$\coprod_{i=1}^{n} x_i = \max(x_1, x_2, \ldots, x_n) = 1 - \prod_{i=1}^{n}(1 - x_i) = 0$$

$$\Rightarrow \quad x_1 = x_2 = \cdots = x_n = 0 \Rightarrow \phi(\bar{x}) = 0.$$

Thus the procedure of a coherent system is bounded.
Since,

$$\min(x_1, x_2, \ldots, x_n) = 1 \Rightarrow \phi(x_1, x_2, \ldots, x_n) = 1$$

$\Rightarrow$ a series system and

$$\max(x_1, x_2, \ldots, x_n) = 1 \Rightarrow \phi(x_1, x_2, \ldots, x_n) = 0$$

$\Rightarrow$ a parallel system, therefore, the bounds of a coherent system is bounded between the bounds of a series and parallel system.

Notation:

$$\begin{aligned} \min(\bar{x}, \bar{y}) &= \bar{x} \vee \bar{y} = (x_1 \vee y_1, x_2 \vee y_2, \ldots, x_n \vee y_n) \\ &= 1 - \prod_{i=1}^{n}(1 - x_i y_i). \end{aligned}$$

Theorem 1 *Let ϕ be a coherent structure.*

Then, (*i*) $\phi(\bar{x} \vee \bar{y}) \geq \phi(\bar{x}) \vee \phi(\bar{y})$
(*ii*) $\phi(\bar{x} \cdot \bar{y}) \leq \phi(\bar{x}) \cdot \phi(\bar{y})$.

Proof of (*i*).
L.H.S. is $\phi(\bar{x} \vee \bar{y}) = \phi(x_1 \vee y_1, x_2 \vee y_2, \ldots, x_n \vee y_n)$.
Now

$$x_1 \vee y_1 = \max(x_1, y_1) \geq x_1 \cdot \tag{5.5.2}$$

$$\text{and } x_2 \vee y_2 = \max(x_2, y_2) \geq x_2 \tag{5.5.3}$$

$$\cdots \quad \cdots \quad \cdots$$

$$\text{and } x_n \vee y_n = \max(x_n, y_n) \geq x_n. \tag{5.5.4}$$

Similarly,

$$x_1 \vee y_1 = \max(x_1, y_1) \geq y_1 \tag{5.5.5}$$

$$\text{and } x_2 \vee y_2 = \max(x_2, y_2) \geq y_2. \tag{5.5.6}$$

$$\cdots \quad \cdots \quad \cdots$$

Therefore,

$$\begin{aligned} \phi(\bar{x} \vee \bar{y}) &= \phi(x_1 \vee y_1, x_2 \vee y_2, \ldots, x_n \vee y_n) \\ &= \phi(t_1 \geq x_1, t_2 \geq x_2, \ldots, t_n \geq x_n) \\ &\geq \phi(x_1, x_2, \ldots, x_n) = \phi(\bar{x}), \end{aligned} \tag{5.5.7}$$

since ϕ is non-decreasing for its arguments.

Similarly,

$$\phi(\bar{x} \vee \bar{y}) \geq \phi(y_1, y_2, \ldots, y_n) = \phi(\bar{y}). \tag{5.5.8}$$

Finally,

$$\begin{aligned}\phi(\bar{x} \vee \bar{y}) &\geq \max(\phi(\bar{x}), \phi(\bar{y})) \\ &= \phi(\bar{x}) \vee \phi(\bar{y})\end{aligned}$$

and

$$x_n \vee y_n = \max(x_n, y_n) \geq y_n.$$

Proof of (ii).

$$\begin{aligned}\phi(\bar{x} \cdot \bar{y}) &= \phi(x_1 y_1, x_2 y_2, \ldots, x_n y_n) \\ &= \phi(\min(x_1, y_1), \min(x_2, y_2), \ldots, \min(x_n, y_n)) \\ &= \phi(t_1 \leq x_1, t_2 \leq x_2, \ldots, t_n \leq x_n) \\ &\leq \phi(x_1, x_2, x_3, \ldots, x_n) = \phi(\bar{x}).\end{aligned}$$

Precisely, in the same way

$$\phi(\bar{x} \cdot \bar{y}) \leq \phi(\bar{y}).$$

Therefore,

$$\phi(\bar{x} \cdot \bar{y}) \leq \min(\phi(\bar{x}), \phi(\bar{y})) = \phi(\bar{x}) \cdot \phi(\bar{y}).$$

5.6 Coherent Systems in Terms of Paths and Cuts

Let $C = (1, 2, \ldots, n)$ be the set of all components and $X = (x_1, x_2, \ldots, x_n)$ denote the states of the sets of components.

We define,

$$\begin{aligned}C_0(x) &= [i \mid x_i = 0] \\ \text{and } C_1(x) &= [i \mid x_i = 1].\end{aligned}$$

Assume that the structure (C, ϕ) is coherent. That means that none of the components in C is irrelevant and $\phi(X)$ is a monotonically non-decreasing function of X, i.e., of $x_1, x_2, \ldots, x_n$.

Definition 1

(*i*) *A path vector is a vector X such that $\phi(X) = 1$.*

(*ii*) *The corresponding path set is again denoted by $C_1(x) = [i \mid x_i = 1]$.*

(*iii*) *A minimal path vector is a path vector X such that for any other vector $Y < X$, $\phi(Y) = 0$, where $Y < X \Rightarrow y_i \leq x_i$ for $i = 1, 2, \ldots, n$, excepting for some i, where strict inequality holds.*

(*iv*) *The corresponding minimal path set is denoted by $C_1(x) = [i \mid x_i = 1]$. Physically, a minimum path set is obtained by taking a minimal set of components whose functioning ensures the functioning of the system.*

(*v*) *A cut vector is a vector such that $\phi(X) = 0$.*

(vi) *The corresponding cut set is* $C_0(x) = [i \mid x_i = 0]$.

(vii) *A minimal cut vector is a cut vector* X *such that for any* $Y > X \Rightarrow \phi(Y) = 1$.

(viii) *The corresponding minimal cut set is* $C_0(x) = [i \mid x_i = 0]$. *Physically, a minimal cut set is a minimal set of components whose failure ensures the failure causes the system to fail. It may be noted that if* X *is a path vector then for any* $Y \geq X (\Rightarrow y_i \geq x_i) \phi(Y) = 1$. *Similarly, if* X *is a cut vector, then for any* $Y \leq X (\Rightarrow y_i < x_i)$, $\phi(Y) = 0$.

5.6.1 Bridge Structure

We illustrate these concepts by considering a Bridge Structure shown below in the following figure:

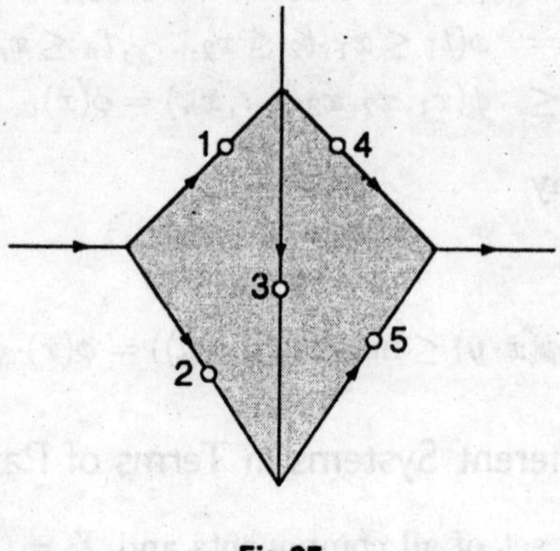

Fig 25

From the above figure the four minimal paths are given by $P_1 = (1,4)$, $P_2 = (2,5)$, $P_3 = (1,3,5)$, $P_4 = (2,3,4)$ since $\phi(x_1, x_4) = \phi(x_2, x_5) = \phi(x_1, x_3, x_5) = \phi(x_2, x_3, x_4) = 1$. This implies that $\phi(y_1, y_4) = 0$, $\phi(y_2, y_5) = 0$, $\phi(y_1, y_3, y_5) = 0$ and $\phi(y_2, y_3, y_4) = 0$ for $y_i < x_i$. Similarly, minimal cut sets are given by $K_1 = (1,2)$, $K_2 = (4,5)$, $K_3 = (1,3,5)$ and $K_4 = (2,3,4)$; since $\phi(y_1, y_2) = \phi(y_4, y_5) = \phi(y_1, y_3, y_5) = \phi(y_2, y_3, y_4) = 1$ implies that $\phi(x_1, x_2) = \phi(x_4, x_5) = \phi(x_1, x_3, x_5) = \phi(x_2, x_3, x_4) = 0$ for $y_i \geq x_i$.

5.6.2 Representation of Minimal Path and Minimal Cut by Binary Functions

Let us define binary function ρ_j corresponding to jth path P_j as $\rho_j(x) = \prod_{i \in P_j} x_i$ given by

$$\rho_j(x) = \begin{cases} 1, & \text{if all } x_i \in P_j \\ 0, & \text{otherwise.} \end{cases}$$

ρ_j is defined to be minimal path series structure. Clearly, ρ_j is the structure function of a series arrangement of the components of the jth path.

In this case,

$$\phi(X) = \prod_{j=1}^{p} \rho_j(x) \tag{5.6.1}$$

assuming p to be the number of components; also since the structure function is unity if and only if at lest one of the path structures is working.

$$\text{or, } \phi(X) = 1 - \prod_{j=1}^{p} (1 - \rho_j(x)). \tag{5.6.2}$$

Therefore, we can write

$$\phi(X) = \coprod_{j=1}^{p} \prod_{i \in P_j} x_i = 1 - \coprod_{j=1}^{p} \left(1 - \prod_{i \in P_j} x_i\right). \tag{5.6.3}$$

Similarly, we associate a binary function corresponding to a minimal cut set k_j of a coherent structure as

$$k_j(x) = \coprod_{i \in k_j} x_i \tag{5.6.4}$$

which is related to $\phi(x)$ as

$$\phi(X) = \prod_{j=1}^{k} k_j(x) \tag{5.6.5}$$

assuming k number of components. Therefore,

$$\phi(X) = \prod_{j=1}^{k} \coprod_{i \in k_j} x_i \tag{5.6.6}$$

takes the value zero if all the components in the jth minimal set fail and one otherwise. Since $\phi(x) = 0$ if all the components fail and 1, otherwise. Therefore, the above structure is a series of arrangement of the minimal cut parallel structures.

5.6.3 Representation of a Bridge Structure as Parallel-series/Series-parallel structure

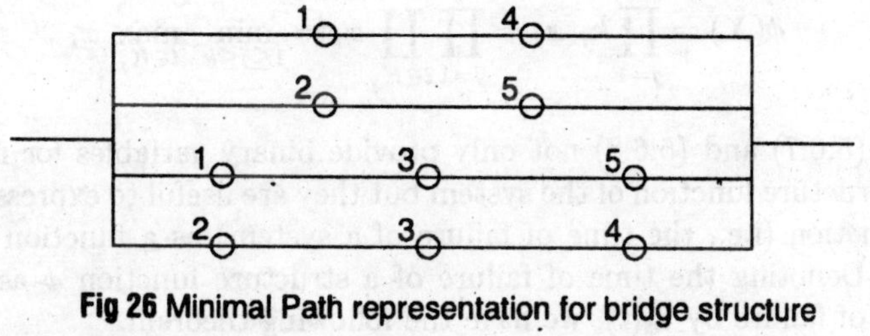

Fig 26 Minimal Path representation for bridge structure

Using (5.6.1), from the above figure we have $\rho_1 = x_1x_4$, $\rho_2 = x_2x_3$, $\rho_3 = x_1x_3x_5$ and $\rho_4 = x_2x_3x_4$ and from (5.6.2)

$$\phi(X) = 1 - (1 - (x_1x_4))(1 - (x_2x_3))(1 - (x_1x_3x_5))(1 - (x_2x_3x_4)).$$

Similarly, using (5.6.4) and (5.6.5), we have

$k_1 = 1 - (1 - x_1)\ (1 - x_2)$, $k_2 = 1 - (1 - x_4)\ (1 - x_5)$, $k_3 = 1 - (1 - x_1)\ (1 - x_3)\ (1 - x_5)$ and $k_4 = 1 - (1 - x_2)\ (1 - x_3)(1 - x_5)$ and $\phi = (k_1) \cdot (k_2) \cdot (k_3) \cdot (k_4)$

$\phi(X) = [1-(1-x_1)(1-x_2)]\,[1-(1-x_4)(1-x_5)]\,[1-(1-x_1)\,(1-x_3)\,(1-x_5)]\,[1-(1-x_2)\,(1-x_3)\,(1-x_5)]$.

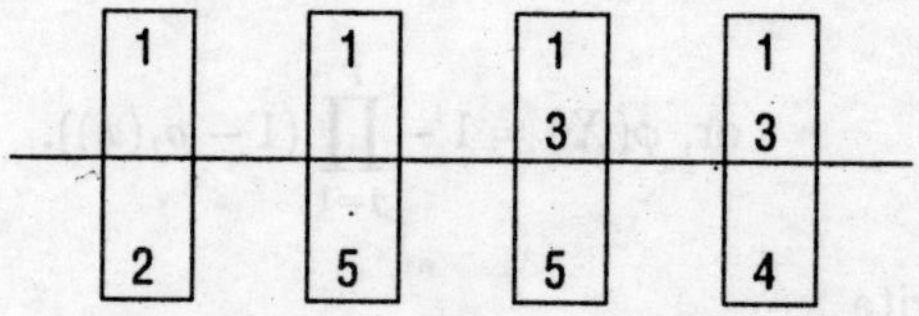

. Fig 27

5.6.4 Time to failure of a system based on the failure of components

Hitherto, we had been considering the failure of a system at a particular point of time based on the failure of components. In the present article, we shall consider systems of operating components over time. Components operate and then fail at a particular point of time ultimately failure of the system . We do not assume provision of the repair of the components. The results (5.6.1) and (5.6.2) can now be extended from binary values to the real positive values.

We introduced the notation for the binary variable, viz.,

$$\prod_{i=1}^{p} x_i = \min(x_1, x_2, \ldots, x_p)$$

$$\text{and} \coprod_{i=1}^{p} x_i = 1 - \prod_{i=1}^{p} x_i = \max(x_1, x_2, \ldots, x_p).$$

Also, (5.6.1) gives

$$\phi(X) = \coprod_{j=1}^{p} \rho_j(x) = \coprod_{j=1}^{p} \prod_{i\in P_j} x_i = \max_{i\le j\le p}\ \min_{i\le\rho_j} x_i. \tag{5.6.7}$$

Also, from (5.6.5)

$$\phi(X) = \prod_{j=1}^{k} k_j(x) = \prod_{j=1}^{k} \coprod_{i\in K_j} x_i = \min_{1\le j\le k}\ \max_{i\in K_j} x_i \tag{5.6.8}$$

The relations (5.6.7) and (5.6.8) not only provide binary variables for minimal paths and cuts in terms of structure function of the system but they are useful to express the time of failure of a structure function (i.e., the time of failure of a system) as a function of the component times of failure. Denoting the time of failure of a structure function ϕ as a function of the component times of failure by $\tau_\phi(t)$, we have the following theorem.

Theorem 2 *Given ϕ to be a coherent structure having minimal path sets $P_1, P_2, \ldots, P_k$ and minimal cut sets $K_1, K_2, \ldots, K_k$, then*

$$\max_{1\le j\le p}\ \min_{i\in P_j} t_i = \tau_\phi(t) = \min_{t\le j\le k}\ \max_{i\in K_j} t_i. \tag{5.6.9}$$

Proof. The identity on the L.H.S. of (5.6.9) is verified by the following considerations:

(*i*) A coherent system fails when the last minimal path structure fails.

(*ii*) A series structure fails at the time of earliest component failure. Similarly, the identity on the R.H.S. of (5.6.9) is verified by the following considerations, viz.,

(*iii*) A coherent system fails when the first minimal cut structure fails.

(*iv*) A parallel structure fails at the time of the failure of the last component.

5.6.5 Reliability of Coherent Structure

Let us denote

$$\begin{aligned} P(X_i = 1) &= p_i \\ \text{and } P(X_i = 0) &= 1 - p_i \end{aligned}$$

$$E(X_i) = p_i \cdot 1 + 0 \cdot (1 - p_i).$$

Therefore

$$E(X_i) = P(X_i) = p_i.$$

Let

$$\begin{aligned} P[\phi(X) = 1] &= h \\ \text{and } P[\phi(X) = 0] &= 1 - h, \end{aligned}$$

where h is the system reliability of a coherent system.

5.7 Reliability of a Hi-Fi system

We have seen in 5.1 that the structure function of a Hi-Fi system is given by

$$\phi(X) = [1 - (1 - x_1)(1 - x_2)] \cdot x_3[(1 - x_4)(1 - x_5)],$$

where x_1, x_2, x_3, x_4 and x_5 denote the states of FM, Record player, Audio, Speaker I and Speaker II respectively.

Therefore,

$$\begin{aligned} E[\phi(X)] &= E = [1 - (1 - x_1)(1 - x_2)] \cdot x_3[(1 - x_4)(1 - x_5)] \\ &= E[(x_1 + x_2 - x_1x_2) \cdot x_3(x_4 + x_5 - x_4x_5)] \\ &= (p_1 + p_2 - p_1p_2) \cdot p_3(p_4 + p_5 - p_4p_5)]. \end{aligned} \tag{5.7.1}$$

It may be seen that for a series system with 3 identical components $h(p) = p3$ and for 2 out of 3 systems $h(p) = 3p2 + p3$ and for a parallel system of 3 components $h(p) = 1 - (1 - p)3$. Also, the system reliability of k out of n systems

$$h(p) = \sum_{i=k}^{n} \binom{n}{i} p^i (1 - p)^{n-1}. \tag{5.7.2}$$

5.8 A Result

$$E[\phi(X)] = p_i h(1_i, X) + (1 - p_i) h(0_i, p).$$

Proof. We have from the Lemma of (5.4), using the result (5.4.1)

$$\phi(X) = x_i \phi(1_i, X) + (1 - x_i)\phi(0_i, X).$$

Taking expectations on both sides, we have

$$E[\phi(X)] = E(X_i) E[\phi(1_i, E(X))] + (1 - E(X_i)) E[\phi(0_i, E(X))].$$

$E(X_i) = p_i$ and $E(X) = (E(X_1) = p_1, E(X_2) = p_2, \ldots, E(X_n) = p_n) = p.$

$$\Rightarrow h(p) = p_i h(1_i, p) + (1 - p_i) h(0_i, p). \tag{5.8.1}$$

Theorem 3 *Let $h(p)$ be the system reliability of a coherent system, then $h(p)$ is an increasing function of p (i.e., for each p_i; $0 \le p_i \le 1$).*

Proof. We have

$$h(p) = p_i h(1_i, p) + (1 - p_i) h(0_i, p) \text{ from (5.7.1)}$$

$$\therefore \quad \frac{dh(p)}{dp_i} = h(1_i, p) - h(0_i, p) = E[\phi(1_i, p) - \phi(0_i, p)].$$

Since, ϕ is coherent, we have

$$\phi(1_i, p) - \phi(0_i, p) \ge 0 \Rightarrow \frac{dh(p)}{dp_i} \ge 0,$$

i.e., $h(p)$ is increasing for each p_i.

Remarks

The redundancy at the component level is more predominant than redundancy at the system level in coherent structure.

Theorem 4 *Let $h(.)$ be the system reliability of a coherent system. Then*

(*i*) $h(p \cup p') \ge h(p) \cup h(p')$;

(*ii*) $h(p.p') \le h(p) \cdot h(p')$.

Equality holds (*i*) when the system is parallel and the same (*ii*) when the system is in series.

Proof. We have for coherent structure

$$\phi(X \cup X') \ge \phi(X) \cup \phi(X')$$

and $\phi(X \cdot X') \ge \phi(X)\phi(X')$ from Theorem 1.

$$\Rightarrow \quad E[\phi(X \cup X')] \ge E[\phi(X) \cup \phi(X')]$$

$$\Rightarrow \quad E[\phi(X \cup X') - \phi(X) \cup \phi(X')] \ge 0$$

$$\Rightarrow \quad \Sigma\Sigma[\phi(X \cup X') - \phi(X) \cup \phi(X')] P[X = x] P[X' = x'] \ge 0$$

$$\Rightarrow \quad \Sigma\Sigma\phi(X \cup X') P[X = x] P[X' = x'] \ge \Sigma\Sigma\phi(X) \cup \phi(X') P[X = x] P[X' = x']$$

$$\Rightarrow \quad h(p \cup p') \ge h(p) \cup h(p'). \text{ This proves } (i).$$

Similarly, the result (*ii*) follows.

5.9 Computing Exact System Reliability

The technique of obtaining the minimal path and minimal cut of a coherent system shown earlier provides technique for computing the reliability of a system.

We have the jth path of a minimal set

$$\rho_j = \prod_{i \in P_j} x_i$$

and the structure function of a path vector is given by

$$\phi(X) = \coprod_{j=1}^{p} \prod_{i \in P_j} x_i \Rightarrow E[\phi(X)] = h(p) = E\left[\coprod_{j=1}^{p} \prod_{i \in P_j} X_i\right].$$

Similarly, the minimal jth cut path is

$$k_j = \coprod_{i \in K} x_i \Rightarrow \phi(X) = \prod_{j=1}^{k} \coprod_{i \in K_j} x_i \Rightarrow E(\phi(X)) = h(p) = E\left[\prod_{j=1}^{k} \coprod_{i \in K_j} X_i\right].$$

Example 1 *Obtain the system reliability of the following structure.*

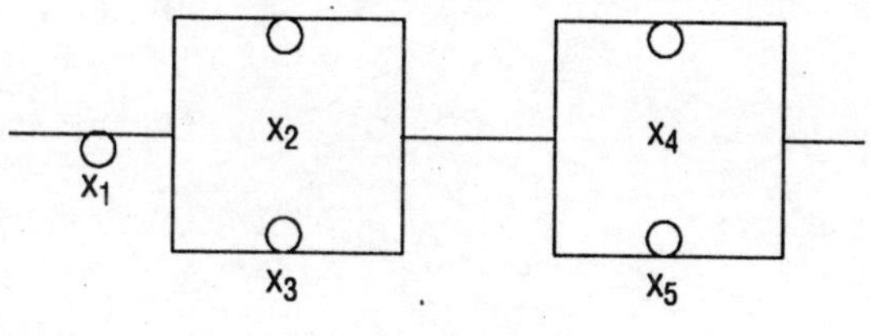

Fig 28

We have,

$$\begin{aligned}
\phi(X) &= x_1 \cap (x_2 \cup x_3) \cap (x_4 \cup x_5) \\
&= x_1 \cdot [1 - (1 - x_2)(1 - x_3)] \cdot [(1 - (1 - x_4))(1 - x_5)] \\
&= x_1 \cdot (x_2 + x_3 - x_2 x_3)(x_4 + x_5 - x_4 x_5) \\
&= x_1 x_2 x_4 + x_1 x_2 x_5 - x_1 x_2 x_4 x_5 + x_1 x_3 x_4 + x_1 x_3 x_5 - x_1 x_3 x_4 x_5 \\
&\quad - x_1 x_2 x_3 x_4 - x_1 x_2 x_3 x_5 + x_1 x_2 x_3 x_4 x_5.
\end{aligned}$$

Hence, the system reliability is

$$\begin{aligned}
h(p) &= E[\phi(X)] = p_1 p_2 p_4 + p_1 p_2 p_5 - p_1 p_2 p_4 p_5 + p_1 p_3 p_4 \\
&\quad + p_1 p_3 p_5 - p_1 p_3 p_4 p_5 - p_1 p_2 p_3 p_4 - p_1 p_2 p_3 p_5 + p_1 p_2 p_3 p_4 p_5.
\end{aligned}$$

Example 2 *Verify the result:* $h(p \cup p') \geq h(p) \cup h(p')$, *where* $p = p'$ *and* $h(p) = h(p')p = p(2p - p^2)$.

We have

$$h(p \cup p') = h[1-(1-(1-p))(1-(1-p'))] = h(p^2) = p^2(2p^2 - p^4).$$

Again

$$h(p) \cup h(p') = [1-(1-p(2p-p^2)]^2 = p^2(2p-p^2)^2. \tag{5.9.1}$$

$$\begin{aligned}
h(p \cup p') &= (p \cup p')[2p \cup p') - (p \cup p')^2 \\
&= [1-(1-p)^2][2(1-(1p)^2) - (1-(1-p)^2)^2] \text{ if } p' = p \\
&= (2p-p^2)[2(2p-p^2) - [1+(1-p)^4 - 2(1-p)^2]. \\
&= (2p-p^2)[2(2p-p^2) - 1 - (1-p)^4 + 2(1+p^2-2p) \\
&= (2p-p^2)[2(2p-p^2) + 1 + 2p^2 - 4p - (1-p)^4] \\
&= (2p-p^2)[4p - 2p^2 + 1 - (1-p)^4 + 2p^2 - 4p] \\
&= (2p-p^2)[1-(1-p)^4] \\
&\geq [1+(1-p)^2][1-(1-p)^2] \quad (\because\ 2p-p^2 \geq 0) \\
&\geq [1+(1-p)^2]^2 \\
&= [2p-p^2]^2.
\end{aligned} \tag{5.9.2}$$

$$h(p \cup p') \geq (2p-p^2)^2 \geq p^2(2p-p^2)^2 = h(p) \cup h(p') \text{ from (5.9.1) and (5.9.2).}$$

Appendix A

Use of IS Variable Sampling Inspection Plans and Tables

A.1 Sampling Plans

Selection of sampling plans by the technique of Indian Standard Institution (ISI) is based on two considerations, viz., the cost of inspection and the quality of protection desired by the producer and consumer. As discussed earlier, the protection provided by a sampling plan to the producer and the consumer is completely determined by its Operating Characteristic curve (OC curve) which provides the probabilities of accepting or rejecting a lot with varying proportion of defective. The power of a sampling plan is decided by the steepness of the OC function. We have seen earlier that steeper is the OC curve the better is the plan. While operating characteristic function gives a complete profile of the protection afforded by the sampling plan it has to be noted that OC function does not give it as a single value (as a function of the OC curve) which may serve as a measure of protection. In view of this, choice of a sampling plan is usually made with reference to certain specified points on the OC curve; as for example, those associated with *Acceptable Quality Level* (AQL), i.e., the maximum per cent defective that, for the purpose of sampling inspection, can be considered as a satisfactory process average). Lot Tolerance Per cent Defective (LTPD), i.e., the percentage of the defective that can be tolerated in only a specified proportion of lots, etc. The AQL/LTPD or such other values may be chosen on the basis of previous data available and also by an agreement between the parties concerned. A comprehensive set of sampling plans classified in terms of AQL is given in the Indian Standard of Sampling Inspection procedures, Part I, for sampling procedures for inspection by attributes (IS: 2500, Part I–2000, ISO 2859–I: 1999) and Part II inspection by variables for per cent defective (IS: 2500, Part II–1965) published by the Bureau of Indian Standards, Manak Bhavan, New Delhi 100 002.

A.2 Classification of IS Sampling Plans

In this section we consider the classification of IS sampling plans based on variables in terms of AQL which includes the following:

- *Variability known plans* (*vide* Table 2, page 16 of IS: 2500, Part II 1965, eighth reprint, September, 2000—reproduced in page 173 of the Appendix).

- *Variability unknown plans, standard deviation method* (Vide Table 3, page 17 of IS: 2500, Part II—reproduced in page 174 of the appendix).

- *Variability unknown plans, range method* (*vide* Table 4, page 18 of IS: 2500, Part II—reproduced in page 175 of the appendix).

A.3 Level of AQL

The choice of a sampling plan for a particular product requires a decision on the quality required. Since AQL plans are used here, it is, therefore, necessary to choose an appropriate level of AQL. While selecting a level of AQL a compromise has to be made between the quality desired and the quality attainable. If AQL is superior to the quality attainable under usual production condition, an excessive amount of product will be rejected. On the other hand, if the level of AQL is inferior to the quality actually attended then an excessive amount of inferior items may be accepted. The value of AQL may be specified by agreement between consumer and producer, giving due weightage, wherever possible, to the past performance in respect of the supply of the product. Sampling plans for certain chosen values of AQL, viz., 0.10, 0.15, 0.25, 0.40, 0.65, 1.00, 1.50, 2.50, 4.00, 6.50, 10.00 are given in Tables 2 to 4 (page 16-18 of IS: 2500, Part II—reproduced in pages 173, 174, 175 of the appendix). In case, the specified AQL is a particular value, other than those above, a suitable value of the same, which is close to the given value may be taken subject to the agreement between the parties concerned.

A.4 Inspection Level

The term *inspection level* is used to denote the relative amount of inspection, one is required to do. Again, the level of inspection needed has to be decided from the quality requirement. A higher inspection level implies relatively more inspection and this entails to lesser risk of accepting lots of quality, worse than the chosen AQL. In other words, the higher the level of inspection, the greater is the protection against low quality lots; but the cost of inspection becomes high. It is, therefore, necessary to make a compromise between the choice of a large sample, which may provide a reliable estimate of the lot quality, and a small, which may considerably cut down the cost of inspection. Such a compromise is possible only by proper choice of the inspection level.

A.5 Description of Table 1

Table 1 given in page 172 overleaf which is auxiliary to other IS tables to be presented subsequently provides five inspection levels with sample size given in codes B (sample size 3), C (sample size 4), D (sample size 5), E (sample size 7), F (sample size 10), G (sample size 15), H (sample size 20 or 25 according as the method is standard deviation or range respectively), J (sample size 25 or 30 according as the method is standard deviation or range respectively), L (sample size 35 or 40 according as the method is standard deviation or range respectively), M (sample size 40 or 50 according as the method is based on standard deviation and range respectively) and finally N (sample size 50 by the method of standard deviation). Again, inspection level I implies the smallest sample size, which minimises the cost of inspection followed by level II, III, IV and V respectively. Thus inspection level gives relatively the largest sample size thereby minimising both the risk of accepting lots with poor qualities and equally rejecting the good lots. But inspection level V achieves the same at a very high cost. However, one has to compromise between a large sample which shows a reliable estimate of the quality of lot and high inspection cost for the same by taking the sample size corresponding to the inspection level IV.

Table 1: Sample Size, Code Letters by Inspection Levels and Sizes of the Lots

Lot Size	Inspection Levels				
	I	II	III	IV	V
1 to 8	B	B	B	B	C
9 to 15	B	B	B	B	E
16 to 25	B	B	B	C	E
26 to 50	B	B	B	D	F
51 to 100	B	B	C	E	G
101 to 150	B	B	D	F	H
151 to 300	B	C	E	G	J
301 to 500	B	D	F	H	K
501 to 1000	C	E	G	J	J
1001 to 3000	D	F	H	K	M
3001 to 10000	E	G	J	L	M
10000 and above	F	H	K	M	N

*IS: 2500, Part II–1965, page 15.

A.6 Lot Acceptability

The acceptability of a lot of materials submitted for inspection is determined by using one of the IS sampling plans described given below, which are associated for specified values of AQL. Depending upon the specification of one-sided or two-sided limits, the acceptability criteria for (*a*) *the variability known,* (*b*) *variability unknown by the method of standard deviation,* (*c*) *variability unknown by the method* of range are given as follows:

A.6.1 Variability known method

- *One-sided specification limits:* On the basis of the given level of AQL and the sample size code letter chosen, the value of the sample size (n) and a suitable factor (k'') will be available from the table 2 [page 15 of IS: 2500, Part II–1965] whose description is presented below. The mean of the n tests ($\bar{x}$) shall then be calculated and the lot will be declared acceptable if:
  - the value of the expression $(\bar{x} + k''\sigma) \leq U$, where U is the upper specification limit which is given or
  - the value of the expression $(\bar{X} - k''\sigma) \geq L$, where L is the lower specification limit which is given.
- *Two-sided specification limits:* The lot shall be declared acceptable if:
  - the value of the expression $\sigma/(U - L) \leq$ the upper limit specified below for the chosen AQL,

– the value of the expression $(\bar{x} + k''\sigma) \leq U$, and

– the value of the expression $(\bar{x} - k''\sigma) \geq L$.

Table 1A: Upper Limit of σ/U–L											
AQL (%)	0.10	0.15	0.25	0.40	0.16	1.00	1.50	2.50	4.00	6.50	10.00
Upper Limit of $\sigma/(U-L)$	0.152	0.158	0.165	0.175	1.84	1.94	0.206	0.223	0.243	0.270	0.304

A.7 A Description of Table 2 [IS: 2599, Part II–1965, page 16]

Title: *Single sampling variables plans for variability known method.*

Given the AQL values (in per cent defective), viz., 0.10, 0.15, 0.25, 0.40, 0.16, 1.00, 1.50, 2.50, 4.00, 6.50 and 10.00 and the sample size code, viz., B, C, D, E, F, G, H, J, K, L, M and N, the table provides n (the sample size) and k'' the multiplying factor with σ (the known standard deviation).

To understand application of tables 1 and 2 let us consider the following example.

Example 1 *The specified minimum yield point for certain steel castings is* 40.0 kg/mm^2.

Suppose lots containing 450 items are brought for inspection. A single sampling variable plan with AQL level of 2.5% and inspection level III is adopted for inspection.

Comment on the acceptance of the lots as per IS plans, given the variability level σ is 1.5 kg/mm^2.

Now the lot of 450 items with inspection level III will have sample size code letter F from Table 1 when the standard deviation is known.

Table 2: Single Sampling Variable Plans for Variability Known Method

Sample Size Code	0.10 n	0.10 k″	0.15 n	0.15 k″	0.25 n	0.25 k″	0.40 n	0.40 k″	0.65 n	0.65 k″	1.00 n	1.00 k″	1.50 n	1.50 k″	2.50 n	2.50 k″	4.00 n	4.00 k″	6.50 n	6.50 k″
B		↓		↓		↓		↓		↓		↓		↓		↓		↓		↓
C		↓		↓		↓		↓		↓	2	1.36	2	1.25	2	1.09	2	0.936	3	0.755
D		↓		↓		↓		↓	2	1.58	2	1.42	2	1.33	3	1.17	3	1.01	3	0.825
E		↓		↓	2	1.94	2	1.81	3	1.69	3	1.56	3	1.44	4	1.28	4	1.11	5	0.919
F		↓	3	2.19	3	2.07	3	1.91	4	1.80	4	1.69	4	1.53	5	1.39	5	1.20	6	0.991
G	4	2.39	4	2.30	4	2.14	5	2.05	5	1.88	6	1.78	6	1.62	7	1.45	8	1.28	9	1.07
H	5	2.46	5	2.34	6	2.23	6	2.08	7	1.95	7	1.80	8	1.68	9	1.49	10	1.31	12	1.11
J	6	2.49	6	2.37	7	2.25	8	2.13	8	1.96	9	1.83	10	1.70	11	1.51	13	1.34	15	1.13
K	7	2.50	7	2.38	8	2.26	9	2.13	10	1.99	11	1.86	12	1.72	13	1.53	15	1.35	18	1.15
L	8	2.54	9	2.45	9	2.29	10	2.16	11	2.01	12	1.88	14	1.75	15	1.56	18	1.38	20	1.17
M	9	2.54	10	2.45	11	2.31	12	2.18	13	2.03	14	1.89	15	1.75	18	1.57	20	1.38	23	1.17
N	11	2.59	11	2.49	13	2.35	14	2.21	16	2.07	17	1.93	19	1.79	22	1.61	25	1.42	29	1.21

All AQL values are in per cent defective.

↓ Use first sampling plan below the arrow, that is, both sample size as well as k″-value.
When sample size equals or exceeds lot size, every lot shall be inspected.

Again the sample size code letter F with AQL level of 2.5% gives the sample size n to be 5, and k'' to be 1.39 from Table 2. This means that a sample size of 5 at random has to be drawn from the lot and their yield points are recorded.

Suppose the yield points for the 5 castings are 43.1, 40.9, 41.7, 44.2 and 41.6. The mean $\bar{x}$ is then 42.3. Also the value of $(\bar{x} - k''\sigma)$ is (42.3–1.39* 1.5) = 40.215. Since this value is greater than the lower specification limit 40.0, the lot should be accepted.

A.8 Description of the Table 3 [IS: 2500, Part II–1965, page 17]

Title: *Single sampling variables plans for variability unknown—standard deviation method.*

- *For one-sided specification limit:* Given the level of AQL and the sample size (vis-a-vis sample size code letter).

  This table gives the value of k', where k' is the multiplying factor of s, and s^2 is unbiased estimator of σ^2. The mean $\bar{x}$ and the standard deviations (with divisor $(n-1)$) shall be calculated from the test results and the lot shall be declared as acceptable if (a) $\bar{x} + k's \leq U$ (when the upper specification limit U is given or $\bar{x} - k's \geq L$ (when the lower specification limit L is given).

- *For two-sided specification limits:* The lot shall be declared as acceptable if (b) the value of $\sigma/U - L \leq$ the maximum value specified in Table 5 [IS: 2500, Part II–1965, page 19] for the particular combination of the chosen AQL and the sample size code letter (c) the value of $(\bar{x} + k's) \leq U$ and (d) the value of $(\bar{x} - k's) \geq L$.

Table 3: Single Sampling Variables Plans for Variability Unknown—Standard Deviation Method

Sample Size Code Letter	Sample Size	Acceptable Quality Level										
		0.10	0.15	0.25	0.40	0.65	1.00	1.50	2.50	4.00	6.50	10.00
		k′	k′	k′	k′	k′	k′	k′	k′	k′	k′	k′
B	3						↓	↓	1.12	0.958	0.765	0.566
C	4					↓	1.45	1.34	1.17	1.01	0.814	0.617
D	5			↓	↓	1.65	1.53	1.40	1.24	1.07	0.874	0.675
E	7		↓	2.00	1.88	1.75	1.62	1.50	1.33	1.15	0.955	0.755
F	10	↓	2.24	2.11	1.98	1.84	1.72	1.58	1.41	1.23	1.03	0.828
G	15	2.42	2.32	2.20	2.06	1.91	1.79	1.65	1.47	1.30	1.09	0.886
H	20	2.47	2.36	2.24	2.11	1.96	1.82	1.69	1.51	1.33	1.12	0.917
J	25	2.50	2.40	2.26	2.14	1.98	1.85	1.72	1.53	1.35	1.14	0.936
K	30	2.51	2.41	2.28	2.15	2.00	1.86	1.73	1.55	1.36	1.15	0.946
L	35	2.54	2.45	2.31	2.18	2.03	1.89	1.76	1.57	1.39	1.18	0.969
M	40	2.55	2.45	2.31	2.18	2.03	1.89	1.76	1.58	1.39	1.18	0.971
N	50	2.60	2.50	2.35	2.22	2.08	1.93	1.80	1.61	1.42	1.21	1.00

All AQL values are in per cent defective.

↓ Use first sampling plan below the arrow, that is, both sample size as well as k'-value.
When sample size equals or exceeds lot size, every item in the lot shall be inspected.

Example 2 (Application of table 3) *The specified maximum resistance of a certain component is* 675 *ohms. Suppose lots containing* 140 *components are submitted for inspection. It is decided to use inspection level* IV, *an* AQL *of* 1.50% *and single sampling variables plan for variability unknown (standard deviation method). Comment on the acceptability of the lot.*

Table 1 gives the sample size code letter corresponding to a lot of 140 components as F. Table 3 gives the sample size (from sample size code letters) as 10 and the factor k' as 1.58. Suppose the resistance of the 10 components selected at random from the lot are 640, 641, 651, 648, 663, 638, 650, 645, 658 and 648.

Then $\bar{x} = 648.2$ and $s = 8.311824$ (with $(n-1)$ in the divisor).

Therefore, $\bar{x} + k's = 648.2 + 1.58 * 8.311824 = 661.3327$.

Since this value is less than the maximum specification limit of 675 ohms, therefore, by A.8 the lot should be accepted.

A.9 Description of Table 4 [IS: 2500, Part II–1965, page 17]

Title: *Single sampling variable plans for variability unknown—Range method. For one-sided specification limit.*

This table like Table 3 gives the values of k (the multiplying factor of the given sample size (*vis-a-vis* sample size code letter) and the level of AQL.

The mean ($\bar{x}$) and the Range R (if the sample size is less than 10) or the mean ($\bar{x}$). And the mean range ($\bar{R}$) (if the sample size is more than 10) shall then be calculated from the result. The lot shall be declared acceptable if: (*a*) $(\bar{x} + kR)$ or $(\bar{x} + k\bar{R}) \leq U$, the upper specification limit (*known*) or (*b*) $(\bar{x} - kR)$ or $(\bar{x} - k\bar{R}) \geq L$, the lower specification limit (*known*).

Table 4: Single Sampling Variables Plans for Variability Unknown—Range Method

Sample Size Code Letter	Sample Size	Acceptable Quality Level										
		0.10	0.15	0.25	0.40	0.65	1.00	1.50	2.50	4.00	6.50	10.00
		k	*k*	*k*	*k*	*k*	*k*	*k*	*k*	*k*	*k*	*k*
B	3	↓	↓	↓	↓	↓	↓	↓	0.587	0.502	0.401	0.296
C	4	↓	↓	↓	↓	↓	0.651	0.598	0.525	0.450	0.364	0.276
D	5	↓	↓	↓	↓	0.663	0.614	0.565	0.498	0.431	0.352	0.272
E	7	↓	↓	0.702	0.659	0.613	0.569	0.525	0.465	0.405	0.336	0.266
F	10	↓	0.916	0.863	0.811	0.755	0.703	0.650	0.579	0.507	0.424	0.341
G	15	0.999	0.958	0.903	0.850	0.792	0.738	0.684	0.610	0.536	0.452	0.368
H	25	1.05	1.01	0.951	0.896	0.835	0.779	0.723	0.647	0.571	0.484	0.398
J	30	1.06	1.02	0.959	0.904	0.843	0.787	0.730	0.654	0.577	0.490	0.403
K	35	1.07	1.02	0.964	0.908	0.848	0.791	0.734	0.658	0.581	0.494	0.406
L	40	1.08	1.04	0.978	0.921	0.860	0.803	0.746	0.668	0.591	0.503	0.415
M	50	1.09	1.05	0.988	0.931	0.893	0.812	0.754	0.676	0.598	0.510	0.421

All AQL values are in per cent defective.

↓ Use first sampling plan below the arrow, that is, both sample size as well *k* value.
When sample size equals or exceeds lot size, every item in the lot shall be inspected.

Example 3 (Application of Table 4) *If in example 2, decision is required to be taken on the basis of range instead of Standard devitation, then examine the acceptability of the lot.*

Table 4 gives the sample size code number F (sample size 10) and the value of k to be 6.50, which means $\bar{x}+k*R = 648.2+6.50*23 = 663.15$. This value being less than the upper limit, the lot should be acceptable by A.9.

A.10 Description of Table 5 [IS: 2599, Part II–1965, page 19]

Title: *The upper limit for the value of* $\frac{s}{U-L}$ *for variability unknown—standard deviation method.*

Two-sided specification limits

Given AQL and sample size code number and the level of AQL, the table provides the upper limit $\frac{s}{U-L}$. The lot shall be declared as acceptable if (a) $\frac{s}{U-L} \leq$ maximum value specified in Table 5 (page 19) and the particular combination of the chosen AQL and the sample size code letter, (b) the value of the expression $(\bar{x}+k's) \leq U$ when the upper specification limit U is given and (c) $(\bar{x}-k's) \geq L$.

Table 5: The Upper Limit for the value of $\frac{s}{U-L}$ for Variability Unknown— Standard Deviation Method

Sample Size Code Letter	Sample Size	Acceptable Quality Level										
		0.10	0.15	0.25	0.40	0.65	1.00	1.50	2.50	4.00	6.50	10.00
B	3	—	—	—	—	—	—	—	0.436	0.453	0.475	0.502
C	4	—	—	—	—	—	0.339	0.353	0.374	0.399	0.432	0.472
D	5	—	—	—	—	0.294	0.308	0.323	0.346	0.372	0.408	0.452
E	7	—	—	0.242	0.253	0.266	0.280	0.295	0.318	0.345	0.381	0.425
F	10	—	0.214	0.224	0.235	0.248	0.261	0.276	0.298	0.324	0.359	0.403
G	15	0.195	0.202	0.211	0.222	0.235	0.248	0.262	0.284	0.309	0.344	0.386
H	20	0.190	0.197	0.206	0.216	0.229	0.242	0.255	0.277	0.302	0.336	0.377
J	25	0.187	0.193	0.203	0.212	0.225	0.238	0.251	0.273	0.297	0.331	0.372
K	30	0.185	0.192	0.201	0.210	0.223	0.236	0.249	0.270	0.295	0.328	0.369
L	35	0.183	0.189	0.198	0.208	0.220	0.232	0.245	0.266	0.291	0.323	0.364
M	40	0.182	0.188	0.198	0.207	0.219	0.232	0.245	0.266	0.290	0.323	0.363
N	50	0.178	0.184	0.194	0.203	0.214	0.227	0.241	0.261	0.284	0.317	0.356

Example 4 *The specified hardness range for certain types of spanners is* 45 *to* 55 HRC. *Lots containing* 2500 *spanners are submitted for inspection. A single sampling variable plan with inspection level V and AQL* 4.00% *is adopted. Given the standard deviation, which is* 2.5 HRC, *comment on the acceptability of the lot.*

Table 1 shows that for 2500 lot size and inspection level V the sample size code letter is *M*. Again, from Table 2 it follows that with AQL level 4.00%, the sample size is 20 and k'' (the multiplying factor) is 1.38.

The upper limit $\sigma/U - L = 0.243$ from Table 1A. Also, $\sigma/U - L = 2.5/(55 - 45) = 0.25 > 0.243$, the upper limit. Therefore, by A.10(a), the lot is straight away rejected without drawing a sample size of 20.

Example 5 *The specified melting range for a certain type of wax is* $60°$ *to* $70°C$. *Lots containing* 250 *containers are submitted for inspection. A single sampling variable plan with variability unknown (standard deviation method) with inspection level* IV *and* AQL 2.5 *per cent is adopted. Comment on the acceptability of the lot.*

Table 1 shows the sample size code letter G when inspection level IV in a lot of size 250 is done. Again, Table 3 gives the value of k' with sample size code letter G and AQL 2.50 as 1.47 and the sample size 15. Finally, Table 5 gives the upper limit of the value $S/(U - L)$ as 0.284.

Suppose we have the following melting point observations in 15 containers as follows:

62.54778	69.36566	74.54139	70.89386	64.43585
62.90680	65.99109	60.15854	63.90070	69.15617
65.93257	66.33906	56.59442	63.71839	89.07339

Then the sample mean $\bar{x} = 63.63704$ and standard deviation $s = 4.602319$

$$\bar{x} - k's = 65.63704 - 1.47 * 4.602319 = 58.87163 < 60$$
$$\bar{x} + k's = 65.63704 + 1.47 * 4.602319 = 72.40245 > 70.$$

Also, $S/(U - L) = 0.4602319 > 0.219$, therefore, the lot has to be rejected by A.10 for failure of satisfaction of either (a), (b) or (c).

A.11 Description of Table 6

Title: *The upper limit for the value of $R/(U - L)$ or $\bar{R}/(U - L)$ for variability unknown.*

For Two-sided Specification limits

Given sample size code letter and AQL the table provides the upper limit of $R/(U - L)$ (for sample size up to 5) or $\bar{R}/(U - L)$ (for sample size > 5) for the particular combination of the chosen AQL and the sample size code letter.

The lot shall be declared as acceptable if (a) $R/(U - L)$ or $\bar{R}/(U - L) \leq$ the maximum value specified in table for the particular combination of the chosen AQL and the sample size code letter, (b) $(\bar{x} + kR)$ or $(\bar{x} + k\bar{R}) \leq U$ and (c) $(\bar{x} - kR)$ or $(\bar{x} - k\bar{R}) \geq L$.

Example 6 *Suppose the observations corresponding to melting points in* 15 *containers* (*Example* 6, Chapter 1) *are as follows:*

65.38757	64.42647	67.09627	67.10095	67.54028
68.98117	67.62568	65.38683	64.32112	63.23981
63.41731	67.61370	68.70459	66.75578	65.23322

Comment on the acceptability of the lot.

Mean $\bar{x} = 66.05538$.

Range of the first 5 tests = $67.54028 - 62.42647 = 5.11381$.

Range of the second 5 tests = $68.98117 - 63.23981 = 5.74136$.

Range of the third 5 tests = $68.70459 - 63.41731 = 5.28728$.

Mean range $(5.11381 + 5.74136 + 5.28728) = 5.380817$.

Applying the condition (a) in A.11, we have $\bar{R}/(U - L) = 0.5380817 < 0.688*$.

Now from table 1 for lot size 250 and inspection level IV the sample size code letter is G and from table 4 the sample code letter G and the AQL 2.50 gives the value of k as 0.610. Finally, from table 6, the upper limit of $\bar{R}/(U - L)$ is 0.688* for sample code letter G and AQL 2.5%.

Applying the condition (b) and (c) in A.11, we have, $\bar{x} + k\bar{R} = 66.05538 + 0.610 * 5.38087 = 69.33771 < 70$, and $\bar{x} - k\bar{R} = 66.05538 - 0.610 * 38087 = 62.77305 > 60$.

Therefore, all the conditions (a), (b), (c) in A.11 for acceptability of the lot being satisfied. the lot should be accepted.

Table 6: The Upper Limit for the value of $\frac{R}{U-L}$ or $\frac{\bar{R}}{U-L}$ for Variability Unknown—Range Method
(*Clause* 5.3.3.2 *and Example* 4)

Sample Size Code Letter	Sample Size	Acceptable Quality Level										
		0.10	0.15	0.25	0.40	0.65	1.00	1.50	2.50	4.00	6.50	10.00
B	3	—	—	—	—	—	—	—	0.833	0.865	0.907	0.958
C	4	—	—	—	—	—	0.756	0.788	0.836	0.891	0.965	1.056
D	5	—	—	—	—	0.730	0.764	0.801	0.857	0.923	1.011	1.118
E	7	—	—	0.695	0.727	0.765	0.804	0.846	0.910	0.985	1.086	1.209
F	10	—	0.529	0.553	0.579	0.610	0.642	0.677	0.730	0.793	0.876	0.977
G	15	0.477	0.493	0.517	0.542	0.572	0.602	0.637	0.688*	0.748	0.830	0.928
H	25	0.447	0.463	0.486	0.509	0.537	0.567	0.600	0.649	0.707	0.785	0.879
J	30	0.442	0.457	0.480	0.503	0.531	0.560	0.593	0.642	0.699	0.776	0.870
K	35	0.438	0.454	0.476	0.499	0.527	0.556	0.588	0.637	0.694	0.771	0.864
L	40	0.432	0.447	0.469	0.492	0.519	0.548	0.580	0.628	0.684	0.761	0.852
M	50	0.426	0.441	0.463	0.486	0.503	0.542	0.573	0.621	0.676	0.752	0.843

Appendix B

Mathematical and Statistical Preliminaries

B.1 Introduction

Let $f(t)$ be a real valued function in $[0, \infty)$. Then the Laplace's transformation of $f(t)$ is defined as

$$L[f(t)] = \phi(s) = \int_0^\infty e^{-st} f(t)\, dt, \tag{B.1.1}$$

where $\quad s > 0$, if real
or real part of $\quad s > 0$, if complex.

Given that (i) $f(t)$ is continuous in every interval and (ii) for constants k_1, k_2, and $\theta \ni \forall\, t > k_0 \mid f(t) \mid < k_1 e^{\theta t}$ holds, then

$$L[f(t)] = \phi(s) \text{ exists.} \tag{B.1.2}$$

Corresponding Laplace-Steiltjes transformation is defined as

$$\int_0^\infty e^{-st}\, dF(t).$$

We have

$$\int_t^\infty e^{-st}\, dF(t) - \int_{t+0}^\infty e^{-st}\, dF(t) = [F(t+0) - F(t)]e^{-st}. \tag{B.1.3}$$

Taking $t \to 0$

$$\int_0^\infty e^{-st}\, dF(t) - \int_{0+0}^\infty e^{-st}\, dF(t) = [F(0+0) - F(0)]$$

$$\int_0^\infty e^{-st}\, dF(t) = \int_{0+0}^\infty e^{-st}\, dF(t)[F(0+0) - F(0)]$$

$$\therefore \quad \int_{0+0}^\infty e^{-st}\, dF(t) = \int_0^\infty e^{-st}\, dF(t) - [F(0+0) - F(0)] \tag{B.1.4}$$

$$\int_{0+0}^\infty e^{-st}\, dF(t) \equiv \text{Laplace-Steiltjes transformation (L.S.T.)}$$

$$\int_0^\infty e^{-st}\, dF(t) = \text{Laplace's transformation (L.T.)}$$

$$\text{L.S.T.} = \text{L.T.} - [F(0+0) - F(0)]. \tag{B.1.5}$$

If in particular,

$$[F(0+0) - F(0)] = 0 \Rightarrow \text{L.S.T.} = \text{L.T.} \tag{B.1.6}$$

B.2 Results

- $f(t) = \sum_{i=1}^{\infty} c_i f_1(t);$

  $\phi(s) = L[f(t)] = \sum_{i=1}^{\infty} c_i L[f_i(t)].$

- $g(t) = e^{-st} f(t)$
  $L[g(t)] = \phi(s-t)$ [First shifting theorem of Laplace's transformation]
  where $L[f(t)] = \phi(s)$.

- If $g(t) = f(t-\theta)$ for $t > \theta$
  $= 0$ for $t \leq \theta$

  $\Rightarrow L[g(t)] = e^{-\theta s} L[f(t)] = e^{-\theta s}\phi(t)$

- For $\theta \neq 0$ and $g(t) = f(\theta t)$
  $\Rightarrow L[f(\theta t)] = 1/\theta L[f(t/\theta)] = 1/\theta \phi(s/\theta)$

- $g(t) = \frac{d^n}{dt^n}[f(t)] = f^{(n)}(t)$

  $\Rightarrow L[g(t)] = s^n L[f(t)] - s^{n-1} f(0) - f^{(n-2)} f'(0) \ldots s f^{(n-2)}(0) - f^{(n-1)}(0).$
  The continuity of $f^{(n)}(t)$ as assumed for $\forall\ n = 1, 2, \ldots$

- $L[f(t)] = \phi(s)$

  $\Rightarrow L[f(\theta t)] = L\left[\int_0^t f(u)\,du\right](1/s)\phi(s)$

- $$\begin{aligned} f(t) &= f_1(t)^* f_2(t) \\ &= \int_0^t f_1(\tau) f_2(t-\tau)\,d\tau, \end{aligned}$$

  where * represents convolution of f_1 and f_2 and $L[f_i(t)] = \phi_i(s)$
  $\Rightarrow L[f(t)] = \phi(s) = L[f_1(t)]L[f_2(t)] = \phi_1(s)]\phi_2(s)].$
  In particular,

  $$\begin{aligned} \phi(t) &= f(t)^* f(t)^* \ldots ^* f(t) \\ &= [f(t)]^{*n} \Rightarrow L[\phi(t)] = L[f(t)]^n \end{aligned}$$

  $\phi(t)$ is called the *n-fold convolution* of $f(t)$.

- If $u(t-a) = 0$ for $t < a$
  $= 1$ for $t \geq a$

  $u(t-a)$ is called *Heavside's unit function* (unit step function)

$$\Rightarrow L[u(t-\theta)] = \frac{e^{-at}}{s}.$$

- $$\begin{aligned} L[f(t-a)u(t-a)] &= e^{-at}L[f(t)] \\ &= e^{-as}\phi(s), \end{aligned}$$

  [Second shifting theorem of Laplace's transformation]

- $\delta(t-a)$ (Dirac's delta function) is defined as the limit of
  $\delta_\epsilon(t-a)$ as $\epsilon \to 0$ given by

$$\begin{aligned} \delta_\epsilon(t-a) &= 1/\epsilon, \quad a < t < a+\epsilon \\ &= 0, \quad \text{otherwise} \end{aligned}$$

$$\Rightarrow u(t-a) = \int_t^{-\infty} \delta(t-a)\,dt$$

$$L[\delta(t-a)] = e^{-at},\ a \geq 0.$$

Chart of Some Fundamental Laplace's Transformation

$f(t)$	$L[f(t)] = \phi(s)$
1	$1/s$
e^{at}	$1/(s-a)$
$\frac{t^{n-1}}{(n-1)!}$ for positive integer n	$1/s^n$
$\frac{e^{tat}t^{n-1}}{(n-1)!}$	$1/(s-a)^n$
$(1/a)\sin at$	$1/(s^2+a^2)$
$\cos at$	$s/(s^2+a^2)$
$(1/a)\sinh at$	$1/(s^2-a^2)$
$\cosh at$	$s/(s^2-a^2)$
$t^n f(t)$	$(-1)^n \frac{d^n}{ds^n} L(f(t))$
$(1/t)f(t)$	$\int_0^\infty L[f(t)]ds$
$\int_{n\pi}^{t} \frac{e^{-a(u-n\pi)}(u-n\pi)^n du}{\Gamma(n+1)}$	$\frac{1}{s(a+s)^{n+1}} e^{-n\pi s}$

B.3 A Result of Differential under the Sign of Integration

Under some regularity condition we have

$$\frac{d}{dt}\int_{f_1(t)}^{f_2(t)} u(x,t)\,dx = \int_{f_1(t)}^{f_2(t)} \frac{d}{dt}u(x,t)\,dx + \frac{df_2(t)}{dt}u[f_2(t),t] - \frac{df_1(t)}{dt}u[f_1(t),t]. \quad \text{(B.3.1)}$$

B.3.1 Some results of renewal theory

Let $S_n = X_1 + X_2 + \cdots + X_n$, where X_is are waiting time for $(i-1)$-th renewal to i-th one. N_t represents the number of renewals in $(0,t]$

$$E(N_t) = U(t) = u(t) \text{ renewal function}$$

$$\frac{dU(t)}{dt} = \text{renewal density.}$$

Given $f_n(t) = f_1(t)^*[f(t)]^{*(n-1)}$, where $f_1(t)$ is the interval density of the first renewal and $f(t)$ is the common interval density of the renewals of all orders $n = 2, 3, \ldots$; $[f(t)]^{*(n-1)}$ is the $(n-1)$-fold convolution of $f(t)$

$$Lf_n(t) = L[f_1(t)]\Big\{L[f(t)]\Big\}^{n-1}.$$

Forward renewal equation

$$U(t) = F_1(t) + \int_0^t U(t-\tau)f(\tau)\,d\tau$$

$$u(t) = f_1(t) + \int_0^t u(t-\tau)f(\tau)\,d\tau,$$

where $U(.), u(.), F(.)$ and $f(.)$ are the Renewal functions, Renewal density, Cumulative Distribution Function and Interval Density Function respectively.

B.4 Asymptotic Renewal Theorem

$$\frac{N_t}{1} \xrightarrow{a.s.} \frac{t}{\mu} \Rightarrow \frac{N_t}{t} \xrightarrow{P} \frac{1}{\mu}, \quad \text{(B.4.1)}$$

where $E(X_i) = \mu$, var $(X_i) = \sigma^2$, X_i, the time from $(i-1)$-th to i-th renewals are independent identically distributed random variables for $i = 1, 2, 3, \ldots$

$$\begin{aligned} N_t &= \text{Number of renewals in } (0,t] \\ U(t) &= \text{Average number of renewals in } (0,t] \\ &= \frac{t}{\mu} + \frac{\sigma^2\mu^2}{2\mu^2} + 0(1). \end{aligned} \quad \text{(B.4.2)}$$

B.5 Probability of a Joint Renewal at $x_1, x_2; x_1 < x_2$ denoted by $f(x_2, x_1)$

$$f(x_2, x_1) = f(x_2)\phi(x_1),$$

where $\phi(x_1)$ is the Hazard function at x_1.

We have, $\phi(x) = \dfrac{f(x)}{F(x)}$.

Proof. We have

$$\begin{aligned} P[X_1 < x_1, X_2 < x_2] &= \int_0^{x_1} \frac{F(x_2) - F(x)}{1 - F(x)}\, dF(x) \\ &= \int_0^{x_1} \frac{F(x_2)}{\bar{F}(x)}\, dF(x) - \int_0^{x_1} \frac{F(x)}{\bar{F}(x)}\, dF(x), \text{ where } \bar{F}(x) = 1 - F(x) \\ &= F(x_2) \int_0^{x_1} \frac{dF(x)}{\bar{F}(x)} + F(x_1) + \log[1 - F(x_1)] \end{aligned}$$

$$F(x_1, x_2) = F(x_2)\Phi(x_1) + F(x_1) - \int_0^{x_2} \phi(x)\, dx,$$

where $\int_0^{x_1} \phi(x)\, dx = \Phi(x_1)$ and $\phi(x) = \dfrac{f(x)}{\bar{F}(x)}$

$$\Rightarrow F(x_1, x_2) = F(x_2)\Phi(x_1) + F(x_1) - \Phi(x_2)$$

$$\frac{\partial^2 F(x_1, x_2)}{\partial x_1 \partial x_2} = f(x_1, x_2) = f(x_2)\phi(x_1). \tag{B.5.1}$$

B.6 Time-Dependent Poisson Process

$$P[N(t) = n \mid \lambda(t)] = \frac{e^{\int_0^t \lambda(\tau)\, d\tau} \left[\int_0^t \lambda(\tau)\, d\tau\right]^n}{n!} \tag{B.6.1}$$

$$\begin{aligned} \lambda(\tau) = \lambda \quad & \Rightarrow \text{ Poisson process} \\ = \lambda t^{\alpha-1} \quad & \Rightarrow \text{ Weibull process} \\ \alpha > 1 \quad & \Rightarrow \text{ Increasing Failure Rate (IFR)} \\ \alpha < 1 \quad & \Rightarrow \text{ Decreasing Failure Rate} \\ \alpha = 1 \quad & \Rightarrow \text{ Constant Failure Rate} \end{aligned}$$

For Weibull Process,

$$P[N(t) = n] = e^{-\lambda t\alpha/\alpha} \frac{(\lambda t^\alpha)^n}{n!} \tag{B.6.2}$$

$$\text{Hazard rate} = \lambda t^{\alpha-1} \tag{B.6.3}$$

$$F(t) = 1 - e^{-\lambda t\alpha/\alpha} \tag{B.6.4}$$

$$f(t) = e^{-\lambda t\alpha/\alpha} \lambda t^{\alpha-1}. \tag{B.6.5}$$

Bibliography

[1] Barlow, R.E. and Hunter, L.C. (1961): *Reliability Analysis of a One-Unit System*—Operation Research 9(2), pp 200–208.

[2] Barlow, R.E. and Proschan Frank (1970): *Statistical Theory of Reliability and Life Testing*—Holt, Rinchart and Winston Inc., New York, Chicago, Sanfransisco, Atlanta, Dallas, Montreal, Toronto, London, Sydney.

[3] Barnard, G.A. (1959): *Control Chart and Stochastic Process*—J.R. Statistical Society. Series B. Vol. 21, pp. 239–71.

[4] Bhat U Narayan (1972): *Elements of Applied Stochastic Process*—John Wiley & Sons Inc. New York, Chickestu, Brisbane, Toronto.

[5] Biswas, S. (1980): *On the Extension of Some Results of Counter Models with Poisson Inputs and Their Applications*-Journal of Indian Statistical Association, Vol. 18, pp 45–53.

[6] Biswas, S. (1988): *Stochastic Processes in Demography and Applications*-Wiley Eastern/John Wiley & Sons New York, Chickestu, Brisbane, Toronto, Singapore.

[7] Biswas, S. and Pachal, T.K. (1982): *On the Applications of Palm Probability for Obtaining Inter-Arrival Time Distribution in Weighted Poisson Process*-Calcutta Statistical Association Bulletin, Vol. 32, March-Jun, No. 125–126, pp 111–15.

[8] Biswas, S. Nautial, B.L. and Tyagi, R.N.S. (1983): *Some Results of Stochastic Process Associated with Geiger-Mutter Counter Model*-Sankhya, Series B, Vol. 45, Part 2, pp 271–283.

[9] Champernowne, D.G. (1953): *The Economics of Sequential Sampling for Defectives*—Applied Statistics, Vol. 2, pp 118–130.

[10] Chiang, C.L. (1968): *Introduction to Stochastic Process in Biostatistics*—John Wiley and Sons, New York.

[11] Columbia University Research Group: *Sampling Inspection.*

[12] Cox, D.R. (1960): *Serial Sampling Acceptance Scheme Derived from Bayes Theorem*-Technometrics, Vol. 2, pp 353–60.

[13] Cox, D.R. and Isham, V. (1980): *Point Process*-Chapman Hall, London.

[14] Cox, D.R. and Oakes, D. (1983): *Analysis of Survival Data*-Chapman Hall, London, New York.

[15] Desmond, D.S. (1954): *Quality Control on the Setting of Voltage Regulators*-Applied Statistics, Vol. 3, pp 65–73.

[16] Dodge, H.F. (1943): *A Sampling Inspection Plan for Continuous Production*-Annals of Mathematical Statistics, Vol. 14, pp 269–79.

[17] Dodge, H.E. and Torry, M.N. (1951): *Additional Continuous Sampling Inspection Plan*—Industrial Quality Control, Vol. 7, pp 5–9.

[18] Duncan, A.S. (1956): *The Economic Design of X-Charts Used to Maintain Current Control of a Process*-Journal of American Statistical Association, Vol. SI, pp 228–42.

[19] Duncan, A.S. (1959): *Quality Control and Industrial Statistics*-Irwin, Homewood, Illinois.

[20] Ewan, W.D. (1960): *When and How to Use CUSUM Chart*—Technometrics, Vol. 5, pp 1–22.

[21] Ewan, W.D. and Kemp, K.W. (1963): *Sampling Inspection of Continuous Process with No Autocorrelation between Successive Results*—Biometrika, Vol. 47, pp 363–80.

[22] Freeman, H.A. Friedman, H., Mosteller, F. and Wallis, W.A. (1948): *Sampling Inspection*-McGraw-Hill, New York.

[23] Gibra (1971): *Economically Optimal Determination of the Parameters of an X-Chart*-Management Science, 17, 635–46.

[24] Girshick, M.A. and Rubin, H. (1953): *A Bayes Approach to a Quality Control Model*-Annals of Mathematical Statistics, Vol. 23, pp 114–25.

[25] Ginedenco, B.V. (1965): *Mathematical Methods in Reliability Theory*-Moscow-English Translation, New York, Academic Press.

[26] Grant, Z. (1946): *Statistical Quality Control*-McGraw-Hill, New York.

[27] Gross, A.J. Clark, V.A. and Liu, V.A. (1971): *Estimation of Survival Parameters when One of the Two Organs must function for Survival*-Biometrics, Vol. 27, pp 36877.

[28] Hald, A. (1981): *Statistical Theory of Sampling Inspection by Attributes*-Academic Press, London, New York, Toronto, Sydney, San Fransisco.

[29] Hill, I.D. (1956): *Modified Control Limits*-Applied Statitics, Vol. 5, pp 12–19.

[30] Horsenell, G. (1954): *The Determination of Single Sample Schemes for Percentage Defectives*—Applied Statistics, Vol. 3, pp 150–58.

[31] Johnson, N.L. (1961): *A Simple Theoretical Approach to Cumulative Sum Charts*-Journal of American Statistical Association, Vol. 56, pp 835–40.

[32] Karlin, S. and Taylor, H.M. (1975): *A First Course in Stochastic Process, Edition II*—Academic Press, New York.

[33] Kemp, K.W. (1958): *Formulae for Calculating the Operating Characteristic and the Average Sample Number of Some Sequential Trials*—Journal of Royal Statistical Society, Series B, Vol. 20, pp 379–86.

[34] Kemp, K.W. (1961): *The Average Rural Length of the Cumulative Sum Chart when a V-mask is Used*–Journal of Royal Statistical Society, Series B, Vol. 23, pp 149–53.

[35] Kemp, K.W. (1962): *The Use of Cumulative Sum Chart for Sampling Inspection Schemes*—Applied Statistics, Vol. 11, pp 16–31.

[36] Khintchine, A.Y. (I960): *Mathematical Models in the Theory of Queing* (Translated from Russian by D.M. Andrews and H.M. Quenouille)—Charles Griffin and Company.

[37] Lieberman, Q.J. and Resnikoff, G.S. (1995): *Sampling Plans for Inspection by Variables*Journal of American Statistical Association, Vol. 50, pp 457–516.

[38] Lorengen, J.J. and Vance, L.C. (1986): *The Economic Design of Control Charts; A Unified Approach*—Technometrics, Vol. 28, 1, pp 3–10.

[39] Mood, M. (1953): *On the Dependence Sampling Inspection Plans upon Population Distributions*—Annals of Mathematical Statistics, Vol. 14, pp 415–25.

[40] Monlgomery Douglas, C. (1996): Introduction of Statistical Quality Control—John Wiley & Sons Inc., New York.

[41] Owen, D.B. (1967): *Variables Sampling Plans Based on the Normal Distribution*—Technometrics, Vol. 9, pp 417–23.

[42] Page, E.S. (1954): *Continuous Inspection Schemes,* Biometrika, Vol. 41, pp 100–115.

[43] Page, E.S. (1955): *Control Charts with Warning Lines,* Biometrika, Vol. 42, pp 243–57.

[44] Page, E.S. (1961): *Cumulative Sum Charts,* Technometrics, Vol. 3, pp 1–9.

[45] Rau, J.G. (1970): *Optimisation and Probability in Systems Engineering,* D. Van Nostrand.

[46] Roberts, S.W. (1959): *Control Chart Tests Based on Geometric Moving Averages,* Technometrics, Vol. 1, pp 239–50.

[47] Ross, S.M. (1970): *Applied Probability Models with Optimistion Applications*—'Holden-Day'.

[48] Ryabinin, I. (1976): *Reliability of Engineering Systems*—Mir Publishers, Moscow.

[49] Sandier, G.H. (1965): *System Reliability Engineering*—Englewood Cliffs, N.J. Prentice-Hall.

[50] Scheffe and Tukey (1945): *Sequential Analysis of Statistical Data*—Statistics Research Group, Columbia University, Columbia University Press, New York.

[51] Shewart, W.A. (1931): *Economic Control of Quality of Manufactured Product*—D. Van Nostrand Co., New York.

[52] Shewart, W.A. (1939): *Statistical Methods from the Viewpoint of Quality Control*—U.S. Deptt. of Agriculture, Graduate School.

[53] Siegmund (1985): *Sequential Analysis—Tests and Confidence Intervals*—Springer Verlag, New York.

[54] Shooman, M.L. (1968): *Probabilistic Reliability, an Engineering Approach*–McGraw-Hill, New York.

[55] Takacs, L. (1957): *On Certain Sojourn Twin Problems in the Theory of Stochastic Process*—Acta Mathematika Academic Society, Hungary.

[56] Wald, A. (1942): *Asymptotically Shortest Confidence Intervals*—Annuals of Mathematical Statistics, Vol. 13, pp. 127–37.

[57] Wald, A. (1942): *Sequential Analysis*—John Wiley, New York.

[58] Wald, A. and Wolfwitz (1989): *Confidence Limits for Continuous Distribution Functions*—Annals of Mathematical Statistics, Vol. 10, pp 105–18.

[59] Wetherill, G.B. (1960): *Some Remarks on the Bayesian Solution of the Single Sample Inspection Scheme*—Journal of the Royal Statistical Society, Series B, Vol. 25, pp. 1–48.

[60] Wetherill, G.B. (1966): *Sequential Methods in Statistics*, Mathuen, London.

[61] Wetherill, G.B. and Gambling, G.E.G. (1966): *The Decision Theory Approach to Sampling Inspection*—Journal of the Royal Statistical Society, Series Vol. 28, pp 318–416.

[62] Wetherill, G.B. (1977): *Sampling Inspection and Quality Control*—Chapman and Hall, London.

[63] Woolward, R.H. and Gold Smith, P.L. (1964): *Cumulative Sum Technique*—I.C.I. Monograph No. 3, Oliver and Boyd Edinburgh.

[64] Wilks, S.S. (1941): *Determination of Sample size for setting tolerance limits*—Annals of Mathematical Statistics, Vol. 12, pp. 91–96.

Index

Some exciting gifts are up for grabs with this book!!

STATISTICS OF QUALITY CONTROL

Suddhendu Biswas

FREE Gifts Rs. 1000/-

DEAR READERS

By now you must have gone through this book in the entirety. So we invite your assessment on this title with absolute impartiality and candid observations. After all, we publishers blindly depend on your benchmarking. Friends, please fill up the open spaces below with CAPITAL LETTERS. You may add some extra sheets of paper if the space provided is not sufficient for you.

i. What are the important topics of your syllabus which are not adequately covered in this book?

ii. In which chapters and on what topics the treatment of the subject-matter by the author is not satisfactory and why?

iii. Have you come across the typos, factual inaccuracies, inconsistencies, and other forms of mistakes in the book? If yes, then please specify the mistakes and page numbers.

iv. Name top three books on the same subject, which, in your consideration, are better than this book. Could you provide some reasons?

Your 1st Preference: ....

Your 2nd Preference: ....

Your 3rd Preference: ....

v. Any suggestion to improve the quality of illustrations, diagrams etc.?

vi. How did you like the cover design? Any suggestions on colours?

vii. Suggestions for further improvement of production standard:

viii. Who recommended you this book:

☐ Teacher ☐ Friend ☐ Parents ☐ Bookseller ☐ Librarian

ix. Name of the person who has recommended:

Designation: ....

Address: ....

Phone No.: ....

e-mail: ....

x. Name and address of the bookseller from whom you have purchased this book:

....

....

....

xi. Name and address of the library where you would like to find the book:

....

....

xii. Name and address of your institution (Please mention the University or Board):

....

....

xiii. Your name, e-mail, and complete postal address:

....

....

....

xiv. A few other titles published by us that you would like to read:

....

....

The best assessment will be rewarded half-yearly. The award will be in the form of our publications, as decided by the Editorial and Marketing Operations Division, amounting to Rs. 1000/- (Rupees one thousand) only.

....

Please mail the filled up coupon at your earliest to :

Editorial & Marketing Operations Division

NEW CENTRAL BOOK AGENCY (P) Ltd

8/1 Chintamoni Das Lane, Kolkata 700 009

West Bengal, INDIA

Ph. No.: 033 + 22418411, 22418573, 22419094

Fax: 033 + 22419573, 22413852

e-mail: ncbapvtltd@eth.net, ncba@cal2.vsnl.net.in

Website: www.centralgroup.com